Daniel B. St. J. Roosa, Edward Talbot Ely

Ophthalmic and Optic Contributions

Daniel B. St. J. Roosa, Edward Talbot Ely

Ophthalmic and Optic Contributions

ISBN/EAN: 9783337815516

Printed in Europe, USA, Canada, Australia, Japan

Cover: Foto ©berggeist007 / pixelio.de

More available books at **www.hansebooks.com**

OPHTHALMIC

AND

OTIC CONTRIBUTIONS

BY

DANIEL B. ST. JOHN ROOSA, M.D.,

PROFESSOR OF OPHTHALMOLOGY IN THE UNIVERSITY OF THE CITY OF NEW
YORK, AND OF DISEASES OF THE EYE AND EAR IN THE UNIVERSITY
OF VERMONT ; SURGEON TO THE MANHATTAN EYE AND EAR
HOSPITAL, ETC.

AND

EDWARD T. ELY, M.D.,

ASSISTANT TO THE CHAIR OF OPHTHALMOLOGY IN THE UNIVERSITY OF THE
CITY OF NEW YORK ; SURGEON TO CHARITY HOSPITAL ; ASSISTANT SUR-
GEON TO THE MANHATTAN EYE AND EAR HOSPITAL.

Nemo solus Sapit.

NEW YORK
G. P. PUTNAM'S SONS
182 FIFTH AVENUE
1880

Press of
G. P. Putnam's Sons
182 *Fifth Avenue*
New York

PREFACE.

These papers have all been printed in various medical journals, or in the transactions of the American Ophthalmological Society during the last three years. The place and date of their first publication are indicated in the table of contents. They are republished in the present form for two reasons. First, in order that the papers, which nearly all pertain to cognate subjects may be grouped together. Second, because it is believed that thus a wider circulation than was possible in the special journals and the transactions in which they were first printed, may be obtained for them. The writers have taken great pains in discussing the subjects of the papers and the history of the cases, and they hope these topics may be carefully considered by that part of the profession interested in them. They are only regarded as contributions, but it is thought that they may at least assist to form a basis for future research.

New York, *May* 1, 1880.

CONTENTS.

SYPHILITIC DISEASES OF THE INTERNAL EAR. BEING A REVIEW OF SOME RECENT PAPERS UPON THIS SUBJECT.*

By D. B. St. JOHN ROOSA, M.D.

In 1877, I published some cases of what seemed to me to be a disease of the ear, chiefly if not entirely affecting the labyrinth. I even ventured to express the opinion that the principal seat of the lesion in these cases, was probably to be found in the Cochlea, and the name Syphilitic Cochlitis was given to the disease.†

Lest I might be misunderstood as supposing that diseases of the *internal* ear of syphilitic origin, were more common than those of the *middle* ear from the venereal disease, when incorporating this paper into the last edition of my work on the ear, it was stated that "it should not be forgotten that syphilitic affections of the middle ear are perhaps more common than those of the labyrinth." This point is made early in the discussion in order that the statement made in the beginning of one of the papers I am about to review may be met in advance. This statement is, "the prevailing opinion has been that the seat of the lesion is usually in the labyrinth." I do not think that the prevailing opinion of our day has been, that the seat of the lesion of the ear in syphilis is usually in the labyrinth, but rather that it is sometimes, perhaps *often*, found there, and in the cases of "sudden deafness" almost always.

---

* The Relations of the Conducting Mechanism of the Ear to Abnormal Hearing. By Samuel Sexton, M.D. Transactions of the American Otological Society, 1878. Syphilitic Affections of the Ear. By Albert H. Buck, M.D., American Journal of Otology, vol. i, no. 1. The Sudden Deafness of Syphilis, with Cases. By Samuel Sexton, M.D. American Journal of the Medical Sciences, July, 1879.

† *Medical Record.*

In January, 1879, Dr. A. H. Buck published a paper upon
"Syphilitic Affections of the Ear." In this interesting article, Dr.
Buck makes four distinct classes of these cases : 1st, those of the
auricle and auditory canal ; 2, of the middle ear ; 3, of the auditory nerve ; 4, of the middle ear and auditory nerve. At the
close of his article, after having given cases that seemed to belong
to each of these heads, Dr. Buck makes use of the following
qualifying language : "With regard to the cases in which the
auditory nerve, either before or after its entrance into the labyrinth, is the part believed to be principally affected by the constitutional disease, I find again that my material is far too scanty
and too imperfect to justify any special conclusions. In these
cases it is generally assumed that the labyrinth is the seat of the
syphilitic lesion. It is quite possible, however, that the lesion may
be in the middle ear, or in the auditory nerve before it enters the
labyrinth." * * * "Lesions at the oval window, by obstruct-
ing the actions of the stirrup, would likewise be competent to
diminish very materially the power of hearing."

The author further states that lesions in the meatus auditorius
internus and in the minute openings of the *cul-de-sac* of the
meatus and in the bony channels of the modiolus, might produce
changes in the functional capacity of the filaments of the auditory
nerve. The conclusion is then reached, apparently in contradiction of preceding statements, that "we are hardly justified in
using the expression labyrinthine disease, except in those cases
where demonstrable lesions are found in this part of the ear at
the post mortem examination. At the same time, it is difficult to
suggest a better term even for temporary purposes."

In the July number of the *American Journal of the Medical Sciences,* Dr. Samuel Sexton published an article entitled "The Sudden Deafness of Syphilis, with cases," in which, after the opening
statement, that has already been quoted, it is said that "Better
knowledge of the disease (syphilis of the ear) seems to lead to
the conclusion that its chief, if not entire location, is in the
middle ear and its conductive mechanism." This paper is a
sequel of one published in the *Transactions of the American Otological Society* for 1878, by the same author, for there, at the close
of an argument against the idea that the labyrinth is often the
seat of the disease, it is asserted in italics that the "sudden deafness of syphilis has, beyond doubt, its principal seat in the conducting mechanism."

In the *Medical Record* for September 20, 1879. Dr. Sexton's views are reproduced with all the *ex-cathedra* importance of an editorial article, and a review of a book upon the Ear, with the same inspiration and the same dogmatic assertions in regard to this subject, has just appeared.* These are evidences of a systematic attempt to establish, upon the dictum of one author, questions, which, to say the least are not yet se.tled, and about which there is yet room for difference, however strongly it may be asserted that " better knowledge " has overthrown the opinions of those who believe that the labyrinth and the auditory nerve are sometimes the seat of a lesion in syphilis, which causes sudden deafness. I have felt bound to go over the whole subject carefully again, and determine if possible, whether it was time for those of us, who had believed that syphilis does sometimes invade the labyrinth and auditory nerve to change our views. As the best way of reaching the subject, I shall review Dr. Sexton's papers, in as brief a manner as the subject will allow.

After the statements already quoted, the paper continues with a reference to Mr. Saunders' views on this subject.

Saunders was perhaps the best authority of his time, but 'his means of diagnosis were poor, and his cases are not reported with sufficient exactness to make them of any value whatever in this discussion. They simply show that certain persons were relieved of impairment of hearing and tinnitus, after using mercury and sarsaparilla. He is however forcibly struck with the congruity of deafness produced by syphilis, and that which was generally described in his time as nervous deafness. Granting that Saunders' cases of nervous deafness were really such, there is some value in his argument ; but, as I have intimated, we are obliged to reject his testimony because in common with the old authors, until the time of Wilde, the data from which his opinions are formed, are not given with sufficient exactness. So far as they go, they are decidedly against Dr. Sexton's views. Iconoclastic as it may seem, I think it would be better for science if all otological literature up to the time of Wilde were rejected, except so far as it may have a historical value in telling of the failure of the ancients to make exact observations in this department. Of course I am here speaking of therapeutical literature, and not of the anatomical works of Eustachius, Valsalva, Monro, and a few others. It would save a great deal of type if we began our discussions with the

---

* *American Journal of Otology*, vol. i, no. 4, p. 292.

opinions of the men who first observed aural disease in a thoroughly objective manner.

Sir William Wilde is next quoted by Dr. Sexton at some length as an author who sustains his view, or to put it in the writer's own words, he "more nearly approached a solution of the syphilitic affections of the tympanum," which this paper is intended to describe. Again, "but his description comes very near reaching the disease, as it is now believed to exist in the middle ear." Wilde describes "an inflammation of a specific character occurring in the membranes of the tympanal cavity, but chiefly exhibited in the external membrane of the drum."

Dr. Sexton after these words, quotes Wilde as regarding this disease as an affection of the *membrana tympani only*. This is a mistake, for as I long since showed, although Wilde called affections of the middle ear by the name of myringitis, because he believed the *membrana tympani* was chiefly affected, he never was so ignorant of pathology as to suppose as is said here, that this membrane *only* was affected. He knew perfectly well that his chronic myringitis was an affection of the middle ear. Wilde's language is "The disease which I am about to describe is an inflammation of a specific character, occurring in the membranes of the tympanal cavity, but *chiefly exhibited* in the external membrane of the drum."* Whatever Wilde may have thought, he is far from denying that there may be a syphilitic cochlitis, or inflammation of the labyrinth ; he is simply describing a disease of the middle ear. He not only came "very near ; " but he actually described cases of deafness, arising as he thought from lesion of the middle ear only, some twenty-five years before our time. There is no evidence as to what he thought of the possibility of syphilis invading the labyrinth. I do not think that he can fairly be quoted in such a discussion as this, for he seems to have expressed no opinions germane to it.

Since the time of Wilde however, I believe we have been able to classify aural disease more accurately, and consequently that we have been able to recognize some well defined affections of the labyrinth from syphilis and other causes, in a much clearer manner than had before been done.

Schwartze is the next authority quoted, and this is to show that nothing has been done by pathologists in the lesions of the labyrinth. The words quoted from Schwartze are, "what has been

---

* Aural Surgery, London, 1853, p. 261.

done by some in this field of late years with the most earnest
endeavors, is scarcely more than a sad dilletanteism, and has no
value for science."
I think this isolated quotation of Schwartze gives an unfair idea
of his opinions. It is taken from the introduction, and the quali-
fication "some" shows of itself that Schwartze believes that good
work has been done even in the pathology of the labyrinth, for if
we turn to page 156 (of the translation), we find a classification
of the diseases that cause hyperæmia of the labyrinth. They are
"typhus, puerperal fever, acute tuberculosis, etc. It may occur in
acute and chronic inflammations of the tympanum, in meningitis
and congestions, in disturbances of the circulation from various
causes, and also from disturbances in the vaso-motor innervation."
Dr. Sexton also quotes Schwartze as saying that even in the most
acute affections of the tympanum, a simultaneous hyperæmia of
the labyrinth *is* met with only exceptionally. What he actually
says is this : "*From my own anatomical investigations* a simultaneous
affection," etc., "*was* met with only exceptionally." He is very
far from asserting that others may not have met with it. Indeed,
he states in a foot-note to the very sentence quoted by Dr. Sexton,
that Hinton, an author who needs no approval of any otologist,
has met with hyperæmia of the labyrinth forty times. Then again
in Dr. Sexton's quotation of Schwartze's opinion, that an independent and primary, non-traumatic inflammation of the membranous
labyrinth has not yet been anatomically and certainly demonstrated,
the foot-note is omitted in which Biechy and Batissim claim to
have found such an inflammation on dissection, and Schwartze's
opinion that "from clinical observation it is probable that an
acute primary and independent inflammation of the inner ear
occurs not infrequently" is also omitted. But more than all this,
Schwartze, l. c. p. 158, gives a case which was under his observation in 1877, and which afterwards came to dissection, which, to
use Schwartze's own language, "places the existence of a *primary
acute purulent inflammation of the labyrinth* without suppuration of
the middle ear beyond all doubt."
But whatever Schwartze may say, good work has been done by
several men in investigating the lesions of the labyrinth. They
have shown that the membranes of the labyrinth may and do
undergo thickening, atrophy, that hemorrhage may occur into
the labyrinth, that gummy tumors may occur in the meatus
auditorius internus. Granting this, it seems to me almost absurd

to believe that in certain syphilitic cases, like aural lesions may not possibly have occurred. Is it probable, nay, possible, that the labyrinth has been entirely excluded from the effects of syphilis, any more than have the retina, the optic-nerve, or the membranes of the brain?

This is an important point in this discussion, for if I correctly understand the drift of Dr. Sexton's paper, he believes that we should not seek nor expect a lesion in the labyrinth, when we may possibly explain the symptoms by reference to the middle ear, and that we may nearly always thus explain them. But we are not entirely without positive evidence that the labyrinth is invaded in the course of syphilis. Moos reported * a case of secondary syphilis, in which deafness, annoying tinnitus aurium and osteocopic pains in the skull were complained of. The hearing was rapidly destroyed. Death. At the autopsy the right external and middle ear were found intact, sclerosis of the petrous portion of the temporal bone, periostitis in the vestibule and small-celled infiltration of the membranous labyrinth, anchylosis of the stapes to the fenestra ovalis. Trunk of the acusticus unchanged.

Gruber has also reported a similar case.† The argument of the present writer has been, that if we observe symptoms such as were seen in cases where lesions of the internal ear were actually discovered on a post-mortem examination, and if we also find that the ordinary treatment for disease of the middle ear has no effect, while that which has been successful in brain lesions is also successful here, we are justified in assuming that we are probably dealing with an affection of the labyrinth, although we cannot substantiate our opinion by an ocular examination of the parts involved.

We have no quarrel to make with Dr. Sexton's next statement, that a specific character may be "engrafted" upon a simple catarrhal inflammation, but we cannot agree with the inference that when a catarrh is modified by syphilis, sudden and absolute deafness becomes one of its symptoms. We know of no reason why syphilitic exudation in the middle ear any more than a catarrhal one should cause a sudden and absolute deafness. The pressure from a non-specific hyperæmia, catarrh, or thickening will be the same as that from a specific one. It is the situation of the lesion,

---

* *Medical Record*, from *Centralblatt fur Chirurgie*, August 19, (77 ?), from Virchow's Archives.

† *Lehrbuch*, p. 617.

and not its character that determines the amount of deafness. When we know that deafness must be very rarely absolute, unless the central organ be involved, just as we know that blindness can be very rarely absolute, unless the retina and optic nerve are affected, have we not a fair right to conclude that absolute deafness depends upon some lesion of the labyrinth or auditory nerve? In passing, we may call attention to the peculiar nomenclature of the paper under discussion. The author speaks of a (*non-purulent*) *mucous catarrhal inflammation of the middle ear*, by which is meant a catarrhal inflammation.

Now let us turn to the cases which are presented to prove that the sudden deafness of syphilis, is dependent upon disease of the middle ear. In case 1, a man of 42 is admitted to St. Francis Hospital, having had syphilis ten years before. He is weak and dizzy and staggers from side to side, and he has pains over his whole head. He has also facial paralysis on the left side. There is no record of any impairment of hearing for nearly three months, and then " deafness became a feature in both ears," whether suddenly or not, we are not told. Certainly all the evidence thus far makes it more likely that a disease extended from his brain to his labyrinth, than from his auditory canal to his drum-heads. He had had plenty of meningeal symptoms. The deafness is nearly absolute, for it is stated that he cannot hear shouting. The physicians at St. Francis Hospital thought the patient had a brain tumor, and I think our readers will agree that he had some kind of brain disease. He now comes under the care of the N. Y Ear Dispensary, seven months after his admission to St. Francis Hospital. He still has facial paralysis, the uvula is drawn to the right side, and he still staggers. His drum-heads show nothing at all marked, he does not hear the tuning-fork well, not at all on the forehead ; he is placed on mercurial treatment, his drum-head is perforated, no fluid is found, and he is finally discharged. Two months afterward he is a little better as to his walking, but no better as to his hearing. This case is gravely reported as one of sudden deafness, dependent upon disease of the middle ear. The present writer can find no proof of what is claimed in the narrative of the case. What evidence there is leans very strongly in his mind towards disease of the labyrinth.

The second case is that of a man of 21, who had syphilis six months before, and woke up five months after the primary sore, with the discovery that he was very deaf. In a few days he could hear nothing. Later he had great pain in the back of the head which lasted for three weeks, during which time he was dizzy, and he had also dimness of vision. He perhaps hears some sounds in the right ear through a trumpet. The tuning-fork is heard when placed on the bones. Exhaustion of the air from both auditory canals, enables him to hear some words through a speaking-tube. He is put upon an active mercurial course. About three months after he passes from observation. His drum-heads, which were retracted and somewhat opaque, are said to be clearing. He cannot hear any words distinctly, however loudly shouted. Dr. Sexton evidently relies upon the fact that the hearing was very slightly improved at one time by rarefaction and condensation of the air in the external meatus, together with the changes in the drum-heads as proofs that the trouble was entirely in the middle ear.

The present writer prefers to believe that a deafness coming on suddenly, and attended by dizziness and staggering gait as well as frontal headache, is much more likely to have depended upon a lesion of the labyrinth, especially since as seen by the doctor's own notes, all the treatment of a mechanical nature directed toward the middle ear had no effect whatever. Now, if mechanical and structural changes in the tympanic cavity cause all these symptoms of central disease, is it not strange that local and mechanical treatment does so little for their cure. We are told that he was put upon an "active mercurial course." If this means what I fear it does not, that this patient received a thorough inunction treatment, together with iodide of potassium in increasing doses as delineated in my published cases, I would be satisfied that the patient had had every chance of recovery. In my opinion no patient with the symptoms that this one presented, would have been thoroughly treated with such prescriptions as are found recorded in the other cases used to illustrate this article. A lesion of the labyrinth is usually, I think, one of the later manifestations of syphilis. As such it will require a very prompt and energetic treatment to arrest it. Besides, the parts involved are so vital that delay in treatment, or inadequate doses of mercury and potash will allow the disease to go on unchecked until it has caused irreparable damage. And I think it probable, that the diseases of the peripheric parts of the body are not only of themselves apt to run their course with more rapidity, but also to be more quickly influenced by treatment than lesions of the brain and the labyrinth. What this may depend upon, if it be a fact, I cannot say. One of the proofs to my mind of the existence of a lesion of the labyrinth, is the fact that such symptoms as these delineated in Case 2, are only relieved by the most active and persistent "mixed" treatment.

The third case of "sudden deafness," reported in the paper under discussion, is one in which the patient stated that her deafness "came on by degrees, in rather a brief period of time." It was certainly a syphilitic case. She had attacks of dizziness, and she could not hear her own voice always when talking, and she is unable to regulate the pitch. These latter symptoms are those which Dr. Sexton in a previous paper has laid great stress upon as evidence of peripheric trouble, and I am quite willing to concede that they show disease probably of the Eustachian tube, and about the fenestra ovalis. But, this by no means excludes much more important changes in the deeper parts. One drumhead was punctured without any effect, except to increase the noises for a short time. She is put upon iodide of potassium, four gr. every four hours, and in three days she is less dizzy, and can hear her own voice most of the time. The patient then disappears from observation for six months, during which time she is said to have been treated for cerebral syphilis. She had several epileptiform seizures, and she took iodide of potassium in large doses. The hearing power scarcely underwent any change, and here the case ends. Certainly there is a strong suspicion that the blood-vessels of the brain were involved in this case, witness the epilepsy, and if those of the brain, why not of the labyrinth. This is, I think, not a clear case of "sudden deafness" from disease of the middle ear alone.

The fourth and last case quoted may be epitomized as follows : A man of 21 lost the hearing of one ear suddenly and absolutely with vomiting and dizziness. Iodide of potassium was given. Two years after, he lost the hearing of the other ear with the same symptoms and then Dr. Sexton saw him. He was anæmic, and had so much vertigo that he was attended when he was on the street. He had severe frontal headache, the pain extending to the vertex. He does not admit having had syphilis, and no positive evidence on that point is presented. There were evidences of sub-acute inflammation of the lower ends of the canals. He heard his own voice distinctly, but the distinctness varies. He does not hear an outside voice at all. The bone conduction is increased by closing the canals. Very low tones uttered close to his ears are painful. He hears all the notes of a piano up to middle C, after that he only distinguishes a rumbling sound. If this be not a symptom of disease of one part of the labyrinth, then our notions of the physiology of the organ of Corti must be revised. How often all of us have seen patients whose labyrinths have probably been ruined by cerebro-spinal meningitis, only able to hear the low notes of the piano. We have always supposed that they heard these, because the auditory nerve with its feeble powers was only able to perceive notes made up of very few and slow vibrations. There is no proof that this case was syphilitic, but we are willing to believe with Dr. Sexton that it probably was. It certainly was a sudden case, and its very suddenness is one of the arguments to prove that it is really one of disease of the labyrinth. Certain it is, that when labyrinth disease does occur, it is with just such symptoms as these. We can hardly imagine a man becoming suddenly and absolutely blind on account of an opacity of the cornea or lens, but how often does a hemorrhage into the sheath of the optic nerve, a plug in the central artery, or an exudation in the macula, destroy all but a glimmer of what we call sight. Just so in my opinion, it is hard to believe that sudden and absolute deafness attended by vertigo and vomiting, can depend on anything less than an exudation, hemorrhage, embolus or tumor, pressing upon some part of the auditory nerve. That other parts are soon involved, or at least, may soon become involved in such a morbid process, I should never think of denying. But, that peripheric disturbances alone can produce such a combination of symptoms I am not able to admit.

Having reviewed the histories of Dr. Sexton's cases I will pass on to the remarks and conclusions that follow them. The throat symptoms are admitted not to have been prominent "nor were the Eustachian tubes found to be obstructed in any of them."

The writer then goes on to remark that he inclines to the opinion that syphilitic lesions seldom if ever reach the middle ear from the throat. This is a view I cannot share. Both in children and in adults, in congenital and acquired syphilis, have I seen cases in which the hyperæmia and catarrh of the pharynx extended to the middle ear, and why should not a syphilitic catarrh as well as a non-syphilitic one creep up through the Eustachian tubes to the middle ear ? Has syphilis such peculiar methods of extension that its inflammatory products pass around the ordinary channels to attack adjacent organs by a circuitous course ? Dr. Sexton's argument seems to be that the tympanic cavity is not only the favorite and almost exclusive situation for syphilitic lesions of the ear, but that they reach this part through the drum head and auditory canal. If this be true, then the affections caused by syphilis are certainly unique.

Dr. Sexton also states that "we know of no cause which produces such peculiar and decided symptoms of deafness" as syphilis. If, by this extraordinary phrase "symptoms of deafness," vertigo and staggering gait are meant, I think that he is in error, for there are a number of causes, for instance the exudation occuring in the course of cerebro-spinal meningitis, mumps, and hemorrhages which produce symptoms very like those of the cases narrated here. It has long since been shown that even syphilitic iritis has no pathognomonic symptoms, and I have yet to learn that we can determine the specific cause of an attack of sudden deafness, by the symptoms, unattended by a history. Considerable stress is laid upon the discovery that in these cases there is a "pre-existing state of hyperæmia in the drums either from cold, or from a sympathetic irritation associated with some affection of the mouth or throat." If I understand this language, it is a direct contradiction of what Dr. Sexton has already said, in regard to the non-extension of syphilitic aural disease from the mouth and throat to the drums. For what is an inflammation of the drums caused by cold, or a sympathetic irritation associated with an affection of the mouth or throat, but an inflammation extending to the drums by the usual channels from the usual causes? Our author gives away, I think, a part of his case when he makes this admission, for he has just been claiming that the throat symptoms were not prominent, and he has tried to show that the drums were reached through the auditory canal.

We pass over the account of the pathology of syphilitic inflammation of the tympanum, for, while it is probably correct, it is open to the same criticism that has been so often made in regard to lesions of the labyrinth, that is, it is purely theoretical, and not founded on any post-mortem examination.

If Dr. Sexton will not allow those of us who believe from the subjective and objective symptoms, that there is a disease of the labyrinth, even if we are not able always to verify our opinions by an examination on the cadaver, neither can he be permitted to base an argument upon a theory that there is an exudation limited to the conductive apparatus, or as he would have us believe, mainly or, perhaps, wholly in the malleo-incudal joint. We are less willing to do this, since the lesion upon which Dr. Sexton lays so much stress, is one as yet scarcely found by the pathological anatomists. A study of Toynbee's catalogue will furnish the evidence upon this point.

Continuing his argument, Dr. Sexton thinks that the labyrinth in these cases is not greatly involved because the auditory nerve responds fully to the sounds conveyed to it, whether from the patient's own vocal cords or a vibrating tuning-fork placed on the skull. Let us see how the histories of his cases justify the use of the adverb "*fully.*" In the first case, the tuning-fork placed on the teeth is heard best in the right ear, but when it is placed on the vertex and glabella it is not heard at all. Can it be possible, that the doctor considers this a full response to the sounds conveyed to the auditory nerve? There is no account as to how the patient hears his own voice, so no argument can be based upon this case. Now, I think that, if this man had had an affection which was even predominantly one of the middle ear, that is to say if the labyrinth was sound or slightly affected, the vibrations of the tuning-fork would have been heard on any part of the skull, and that its sound would have seemed to him to be very loud ; in other words, it would have been intensified. If experience is worth anything upon this subject, it shows that it is especially in affections of the labyrinth that the tuning-fork is not heard at all on some parts of the skull, while in those of the middle ear alone its sound is always intensified.

Then again, in his first case " the patient cannot hear any voice, however loud,"—this is Dr. Sexton's own statement,—" not even shouting." To repeat my argument in a previous part of this paper, I again state that it is very hard, with the knowledge we now have, to believe, that any rigidity of the ossicles, any hyperæmia of the membrane of the middle ear, any amount of fluid in the tympanic cavity, any stricture of the Eustachian tube, or any combination of these conditions, would make a man so deaf that he could not hear any voice, however loud. I appeal to the judgment of those who have seen much aural disease, whether in their opinion anything but a central affection can cause impairment of hearing to such an extent as this? The analogies that I have already frequently used with regard to the affections of the external portions of the eye, as compared with those of the optic nerve and retina, in causing blindness, may be again recalled.

In Case 2, we find that "words shouted through a trumpet into the left ear are unheard," but the patient fancies that he can hear some sounds when the experiment is made in the right ear. "The tuning-fork is heard when placed on the cranial bones." There is no evidence furnished that he heard it fully, or as middle ear cases usually do, intensified.

Rarefaction of the air causes the second patient to hear some words through a speaking tube. This is one of the points upon which the author relies for his argument, of which I shall speak more fully subsequently, that the middle ear is chiefly affected when in any case, change in the density of the air in the external auditory canal alters the hearing power. On being dismissed from treatment, it is said that "he hears the tuning-fork as before, but he cannot hear words distinctly however loudly shouted through a trumpet." This case, from this part of the evidence, seems to have been a mixed one; that is, one in which there was considerable affection both of the middle and internal ears; but if we do not abandon our notions of naming diseases from the part chiefly affected, we should still class this as predominantly one of the labyrinth.

I have never believed that the affection, which I have denominated cochlitis, involved the cochlea solely, but that it affected that part of the ear predominantly, just as a patient may have severe hyperæmia, and even inflammation of the external auditory canal, quite secondary to the main trouble in the middle ear.

It would be very convenient indeed, if we could separate diseased parts from each other by a line as distinct as that in facial erysipelas, or, to use a geographical comparison, as marked as the separation of Mexico from the United States by the Rio Grande ; but the present writer inclines to the view that to give the exact line of demarcation in disease, is very often impossible.

In the third of Dr. Sexton's cases, the patient was absolutely deaf to all external vocal sounds, but she hears her own voice in talking. Yet, sometimes, even that becomes inaudible. "To-day," quoting Dr. Sexton's words, "she could not hear herself scream." "She hears some letters of the alphabet better than others." When dismissed from treatment, "she hears her own voice in the natural tone." "She hears herself sing, but cannot hear herself whistle." "A vibrating tuning-fork is heard when placed on the teeth and mastoid, but is *not* heard when placed on the vertex." She is absolutely deaf as to the voice of others." Here again the tuning-fork is not fully heard. I can only repeat with reference to this case what I have said with reference to the first, that so far as the power of hearing the tuning-fork and the voice shows anything, it indicates, unmistakably in my opinion, disease of the auditory nerve in some part of its course.

In the fourth case, the patient "hears his own voice distinctly, but the distinctness varies, frequently, for a few moments at a time. He hears absolutely no voice in the left ear, but in the right he hears sound when a metallic bougie is struck on the tuning-fork near his ear." A little while before dismissal it was noted that he could "hear all the notes on a piano up to middle C, but above that letter he can only distinguish a rumbling sound." This fact, it seems to me, indicates that the portion of the labyrinth tuned to high notes was more affected than that tuned to lower notes. When he was dismissed from treatment he could not hear any conversation, even through a trumpet. "Very low tones uttered close to his ear were painful." This symptom of pain from *sound* was long since stated by myself to be, perhaps, an evidence taken with other symptoms, of disease of the labyrinth, and I believe it will be found that only those persons who give evidence of hyperæmia or inflammation of the labyrinth, either primary or secondary, are effected by sounds to any unpleasant degree.

The doctor then proceeds to remark that mobility of the drumhead, pathological changes in the ossicles, especially in the malleo-incudal and the stapeo-incudal joint, or fixation of the stapes in the oval window, are sufficient to account for all the phenomena of audition as described in the four cases which I have just cited. I have already expressed my own opinion, namely, that any or all of these changes are not sufficient to produce absolute deafness, to produce inability to hear certain tones at all, or to cause pain to be experienced when sound is conveyed to the ear. He then remarks, after having entitled his paper, "The Sudden Deafness of Syphilis," and having given four cases to illustrate it, two of which are not sudden at all, that the deafness in these cases is not always sudden ; that is to say, it does not always occur suddenly. He then refers for proof of his view, that the changes in the middle ear are sufficient to account for all the symptoms, to a paper published by him in the Transactions of the Otological Society for 1878.

It is to be expected that the majority of the readers of these *Archives* are familiar with this elaborate paper by Dr. Sexton, but since it has been alluded to in this argument, I am obliged to follow him there and to discuss the points which it contains.

As was stated in the beginning of this review, Dr. Sexton's article in *Hays' Journal* is apparently a sequel to the one published in the Transactions of the Otological Society. The latter is an argument in favor of the conducting as against the perceptive parts of the ear as being the seat of the phenomena of audition and disease, but it contains very few proofs for the correctness of the views advanced.

These are supposed to be found in the narration of the cases of the former paper. I am obliged to select merely those points that bear directly on the question under consideration.

Great stress is laid upon the fact that the membrana tympani is capable of transmitting from 16 to 40,000 vibrations a second to the auditory nerve, and upon the opinion of Edward Weber that the bones of the ear, and the petrous bone, are solid incompressible bodies, and that the fluid of the labyrinth is likewise incompressible, also that the ossicles must be regarded as solid levers which transmit waves of condensation and rarefaction to the fluid of the labyrinth moving it as a whole.

It is argued that because in health there is a very free motion or separation in the joint between the malleus and the incus, when disease has increased the separation, symptoms such as autophony and tinnitus may occur.

The separation of the joints is supposed to result from hyperæmia or inflammation of the drum or pathological changes in the ossicles. Then follows a discussion of double-hearing, so-called. Under autophony Dr. Sexton seems to include double-hearing; not hearing one's own voice naturally, hearing one's own voice as if at a distance down in a well or pit, all of which are referred to affections of the middle ear. It is stated that autophony does not occur when the membrana tympani is absent. It is stated that hearing the ticking of a watch and not hearing ordinary conversation, and the contrary, hearing better in a noise, are explainable by the condition of the malleo-incudal joint or drum-head or both, but no proof is given for this statement. I think I have shown that no adequate explanation has ever been given for the phenomenon of hearing better in the midst of noise.*

It is also stated that the effect produced by inhaling chloroform, ether, nitrite of amyl, or by taking large doses of quinine, is hyperæmia of the ear, and consequently temporary separation of the malleo-incudal joint. It is difficult for me to entertain such an explanation as this. That a man may take a dose of quinine, or inhale chloroform or ether or nitrite of amyl, and thereby separate his malleo-incudal joints by hyperæmia, and not at all affect the labyrinth is to me simply incredible.

But Dr. Sexton seems desirous to exclude the labyrinth from having anything whatever to do with hearing, except in a state of health. Even in disease artificially produced, the labyrinth he seems to believe, is isolated from all its surroundings, and enjoys an immunity that is not shown by any other part of the human body.

It is argued that tinnitus cannot probably have its origin in the incompressible cavity of the inner ear, but it is admitted that a sudden increase of blood in the labyrinth can force the stirrup from its close connection with the other ossicles, and that the return will be attended with sound, and in this way we may account, it is said, for the whistling and whirr of labyrinthine vertigo. In other words, the ossicles must still be held to account for an affection which, according to the writer's own statement, begins as a sudden increase of blood in the labyrinth. Thus having said in one breath, that tinnitus cannot have its origin in the labyrinth, in the next, it is stated that the sudden increase of blood in the labyrinth forces the stirrup outward and the return causes a noise ; *ergo*, the origin was not in the labyrinth.

But we are unable to follow Dr. Sexton through his arguments to prove what seems to be his belief, that almost all the symptoms that are seen in diseased ears, from sudden and absolute deafness to vertigo and tinnitus, are chiefly due to abnormity in the conducting mechanism and especially to " separation of the malleo-incudal joint." They are, as it seems to the present writer, of the kind already quoted.

It should be said, before passing on to a review of Dr. Sexton's conclusions, that one of the four cases, No. 4, which he has presented as proofs that the sudden deafness of syphilis is due to an affection of the middle ear, was published by Dr. Buck in his

---

* Treatise on the Ear, p. 512.

paper, he having seen the case in consultation, as one of disease of the labyrinth.*

In classifying these cases Dr. Buck says, " To the *second class,* finally, belong the following seven cases, in all of which it is fair to assume, from the comparatively normal condition of the middle ear and from the history of the case, that the labyrinth or its immediate vicinity was the seat of the pathological changes that caused the deafness." Although I would not attempt to argue from the weight of authority, for this difference of opinion cannot be settled by any reference to the names of those who advocate one view or the other, I cannot refrain from quoting Dr. Buck's words to show, that Dr. Sexton failed to convince the gentleman whom he called in consultation, that his theories were correct.

We now continue our review of the paper published in the *Journal of Medical Sciences.* Dr. Sexton states that the results already obtained do not warrant a favorable prognosis in the cases of sudden deafness arising from syphilis, and that the chances of success are not good, because the lesion is not ushered in with pain in the ear.

It seems to me, that a lesion which is ushered in by sudden and profound deafness, vertigo and great tinnitus is sufficiently alarming without pain, to invite an early consideration. We find no record in the cases given of the energetic mixed treatment advised, not as Dr. Sexton intimates, "by Wilde and later by Roosa," but advised first by Roosa and never by Wilde, so that we cannot say that the author would not modify his own prognosis, if he would resort to the treatment under which my own cases were benefited.

I will now present the conclusions reached by Dr. Sexton, and discuss them *seriatim.*

CONCLUSIONS. 1. " Syphilitic affections of the ear inducing sudden deafness are of exceptional occurrence."

I make no objection to this conclusion ; fortunately they are of exceptional occurrence.

2. " They would seem to be induced by a pre-existing hyperæmia in the ears, excited by sympathetic relationship, or by an inter-current attack of aural mucous catarrh." †

I confess I do not quite understand the point here : How a

---

* *American Journal of Otology,* vol. i, no. 1.

† According to the ordinary lexicographers, *catarrh* is a discharge from, or an inflammation of, a mucous membrane. To use the term mucous catarrh is certainly not to increase the simplicity or correctness of aural nomenclature.

" pre-existing hyperæmia " is excited by a " sympathetic relation-ship," (a sympathetic relation between what?) I do not know. If this means that a person having syphilis, is more liable to a sudden deafness if he has previously suffered from a hyperæmia of the ear or from nasal catarrh, I think none of us will deny it, but as the conclusion stands I have failed to find it intelligible.

3. " The attacks are characterized by their sudden occurrence, and both ears are usually affected simultaneously, although the contrary sometimes takes place."

My experience has led me to believe that in a certain class of cases, those affecting the labyrinth, the attacks of deafness are sudden, but I have seen other cases of affection of the ear, which seemed to me to be caused or modified by syphilis, where the impairment of hearing came on gradually. As I have already said, in one of Dr. Sexton's four cases the deafness cannot be said to have come on suddenly, and in several of Dr. Buck's cases * the same is true.

4. " The deafness is always very great."

With regard to this I have only to say, that when the deafness is so great as to be nearly or quite absolute with regard to the human voice, I should conclude that there was a primary or secondary lesion of the perceptive apparatus, whatever may have happened to the conducting mechanism.

5. " This syphilitic affection speedily causes a disarrangement of the integrity of the chain of ossicles, most likely at the malleo-incudal joint, probably in some instances of the stapedo-incudal joint, or both of these. The movements of the stapes in the oval window are also likely to be interfered with. The two first mentioned conditions serve to explain the noises in the ears, and the autophony ; the last mentioned condition would increase the anomalies of hearing."

This has been fully discussed in going over the subject. It seems to me not to have been proven, but to rest on assertions which even the author's selected cases do not at all substantiate.

6. " The disease is usually unattended by pain in the ears, it is non-purulent, and its incurability is a characteristic."

The affection causing the sudden deafness of syphilis is certainly usually painless. It is also non-purulent. But I cannot admit, if it

---

* l. c.

be thoroughly treated, at an early period, that its incurability is a characteristic.

7. "The affection does not depend, so far as we know, on anomalies of any portion of the labyrinth, although the latter of course is liable to invasions from syphilis with the nature of which we are as yet unfamiliar."

So far as we know, the *sudden* deafness of syphilis, as shown by clinical and pathological investigations, does depend upon disease of the labyrinth. At least the evidence for the truth of this theory, is much stronger in my opinion, than for the one that it depends upon lesions of the middle ear, or especially upon an affection of the malleo-incudal joint.

If the reader will bear with me a short time longer, I will now tabulate the conclusions which I have reached regarding sudden deafness caused by syphilis and other affections, which I do not think have as yet been overthrown.

It is hardly necessary to say, that I am far from believing that these conclusions are absolutely correct and final. Undoubtedly, whatever may have been done as yet in that way, better knowledge will some day modify them.

1. Very great impairment of hearing, occurring suddenly, and not to be explained by the conditions found in the auditory canal or middle ear, so far as we can examine them, and not relieved at once, by mechanical treatment, whether occurring in the course of syphilis or not, probably depends upon a lesion in the labyrinth or auditory nerve.

2. Absolute or nearly absolute deafness, the inability to hear certain tones, are symptoms of either primary or secondary lesion of the labyrinth.

3. If the tuning-fork be heard very feebly or not at all when placed upon the skull, or if it is heard better through the air than when placed upon the bones, it is probable that there is disease of the labyrinth.

4. Syphilitic diseases of the labyrinth, if vigorously attacked by means of mercury and the iodide of potassium soon after the beginning of the disease, may often be alleviated and sometimes cured.

5. Pathological examinations of the labyrinth although not numerous, have already demonstrated that changes may occur there, which confirm the conclusions that have been formed from clinical investigation.

A NEW AURAL DOUCHE.

DURING the past summer Dr. Charles Fayette Taylor, of New York, suffered for a time from an acute suppuration of the middle ear, and found great relief from the use of warm water by means of the *fountain syringe*. His remarkably inventive mind soon turned his attention to the defects in this method of applying water to the ear for the relief of *pain*, and he invented the douche, of which a figure and description are here given. I cordially commend it to my professional brethren as a valuable means of applying warm water to the auditory canal, drum-head, and tympanic cavity. It is not useful as a *syringe*, but as a *douche*. I think it better in most cases than the fountain syringe. It may be obtained of Messrs. John Reynders & Co., 303 Fourth Avenue, New York.

The Fayette Aural Douche consists of two siphons, so arranged that the flow starts at the same moment in each; and while one siphon conveys the water into the ear the other lifts it gently out, without friction or pressure upon the inflamed tissues.

In the figure, *BC* represents the *ear-piece*, which is made of suitable size and shape. Two holes are bored through it, one lying above the other when it is in its proper position. On each of the two projections at the larger end, a piece of flexible rubber tubing (such as is used for nursing-bottles) about four feet long, is fitted. At the small end of the ear-piece the division between the holes is cut back about one-eighth of an inch, so that placing the finger over this end leaves one continuous passage from the top, *A*, to the bottom, *D*. With the finger over the small end of the ear-piece, as just described, when water is poured into the funnel *A* it will flow directly through both tubes, and come out at the lower end, D, in the drip-vessel. When all the air has thus been excluded and a current established, the funnel *A* is dropped into the basin or pitcher which serves as a reservoir, and a single siphon is formed. The rubber tubes are now compressed by the thumb and finger at E, so as to arrest the flow, the finger is removed from the end BC, and the ear-piece is inserted into the

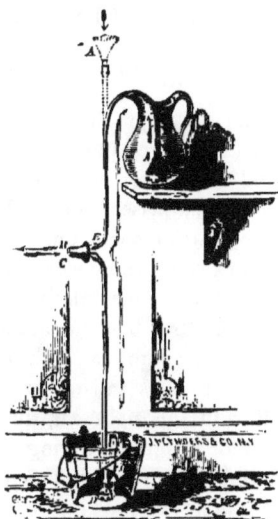

auditory canal : then letting go the tubes at E, a *double* siphon is instantly established, AB conveying the water into the ear, and CD carrying it out by atmospheric pressure. Thus the resistance and pressure, often painful, of the in-coming and out-going currents is avoided, and a small amount of constantly changing water, of any desired temperature, is kept in contact with the auditory canal and drumhead. Any amount of water desired can be used in one continuous bath, without the trouble of refilling the reservoir several times, as is so often required in using the fountain syringe. [D. B. ST. J. R.]

CLINICAL CONTRIBUTIONS TO OTOLOGY.

BY

D. B. St. JOHN ROOSA, M.D.,

AND

EDWARD T. ELY, M.D.

CASE I.—*Loss of hearing from a kiss upon the ear.*

Mrs. H., æt 42, seen through the kindness of Dr. O. B. Douglas. Last winter (1878), her husband came up behind her as she sat reading and kissed her suddenly upon the right ear, taking her completely by surprise. She suffered a great shock and had a roaring in the ear for some time. The incident made her very "nervous" for two or three weeks afterwards. During the past summer she was told by her relatives that she was becoming deaf on the right side. She paid no attention to it until six weeks ago, when she tried her right ear with her watch and found she could not hear it. She gives satisfactory evidence of having heard a whisper well with the right ear during last winter and spring. Has had occasional tinnitus during the past few months after taking cold. Enjoyed music very much formerly, but does not now. The piano-practice of the children at home annoys her. Whistling is particularly disagreeable. All noises disturb her somewhat, so that she has "felt afraid that she was becoming nervous." General health is good. Menstruates regularly. No cardiac trouble detected. Father died of paralysis.

H. D., *R.* $\frac{P}{40}$. *L.* $\frac{40}{40}$.

Tuning-fork on teeth or vertex seemed louder in the left ear. Is slightly intensified in right by plugging, but much more in left. Aerial better than bone-conduction on each side.

The drumheads are about alike and show nothing to account for deafness. Air enters the right drum by both catheter and Politzer's method, but does not alter hearing. All notes of the piano are heard, but she says they do not sound "clear," even with both cars open. Dr. Douglas examined the naso-pharyngeal space and the mouths of the Eustachian tubes and found nothing abnormal.

This seemed to be a case of deafness from affection of the labyrinth, with no apparent cause except the kiss upon the ear. The concussion from the kiss, may have caused the loss of hearing at once : or, as seems more likely, it may have produced changes in the labyrinth, which, in combination with the general nervous shock, served as a foundation for a gradual loss of hearing subsequently.—as, for instance, by some atrophic process.

Mr. Hinton was inclined to think that in all instances of loss of hearing, apparently from slight causes, it might be found that some previous source of injury to the car had existed. He quotes some cases to illustrate that view. He speaks of a concussion sometimes jarring the labyrinth, not into complete paralysis, but into a state of extreme liability to this condition.*

CASE 2.—*Alarming syncope after cleansing ear.*

Mr. G., æt. 40 ; lawyer. Consulted us on February 1, 1879, for a chronic suppuration of the right middle ear, which he had allowed to remain neglected for a long time.

The hearing was : *R.* $\frac{5}{18}$, *L.* $\frac{48}{18}$.

Tuning-fork on teeth heard chiefly in *R. E.*

The ear was syringed with warm water and then cleaned with cotton on cotton-holder, after which the patient complained of feeling faint. He immediately lay down upon the sofa but he did not recover from the syncope as it was expected he would do. He became comatose, and his countenance was livid. His respirations sank to six in the minute and were stertorous. The heart-beat was very feeble and no pulse could be felt at the wrist. He looked as if he were certainly dying. His clothes were loosened as soon as possible, and, by the time this had been done, he opened his eyes and spoke. After this, ammonia was applied to his nostrils and given internally with sherry wine. Electricity was applied also, by Dr. Rockwell, who, together with Drs. Sayre, Bache, Emmet, and Bull, rendered kind assistance. For some time the patient's mind was not perfectly clear, his color was livid and his pulse very feeble ; but finally he became better. The accident happened at 1 P.M., and he remained upon the sofa until 2.45 P.M.

* *Questions of Aural Surgery,* p. 268.

He then went home, complaining of chilly sensations. He did
not look as well, however, as he did before the attack.

We had never before seen any such serious symptoms from
cleansing a tympanic cavity and were at a loss to account for
them. The manipulations were all made with the utmost gentle-
ness. Mr. G. himself said that nothing was done for his ear which
caused the slightest pain or discomfort ; and he attributed the
fainting entirely to " mental influence,"—a sort of dread that he
would possibly be hurt. He had never fainted but once in his life
before, and that was after hearing a friend tell of a surgical operation. He had lately been subjected to overwork and anxiety ; he
had a sallow complexion and gave some symptoms of organic heart
disease. Otherwise, he looked like a strong man.

Mr. G. came after the close of office-hours when there was not
sufficient time for more than a superficial examination of his case.
His history, therefore, was not recorded as fully as it would have
been otherwise. The fainting occurred before the condition of
his middle ear had been determined. After the attack it was not
considered proper to subject him to any further examination, and
he has never been seen again. His case is reported simply to
show what serious consequences may arise from cleansing an ear.

CASE 3.—*Serious syncope from inflation of middle ears by
Politzer's method.*

Miss P., æt 19, came June 24, 1879, complaining of deafness
and "confused feelings" in the right ear. There was a history of
pain and discharge in that ear after scarlet fever, at the age of
2½ years.

The hearing was : $R. \frac{0}{48}$, $L. \frac{48}{48}$. Tuning-fork on teeth heard
better in left ear. Right drumhead cicatricial and hyperæmic.
Left sunken, no light-spot. After inflation by Politzer's method,
this patient had a serious attack of syncope, from which she recovered very slowly. At her next visit she fainted again after a
most gentle inflation through Hinton's tube. The catheter was
not used at either visit.

The improvement in the hearing and in the sensations of the
right ear from inflation made her "feel strange," and this may
have had something to do with the fainting. She had a very
nervous temperament, and was anæmic. She gave the impression
of being too tightly laced, and of being improperly managed generally. Dizziness after inflation is not uncommon : syncope from

inflation by Politzer's method, properly performed, has never before been seen in the writer's experience.

CASE 4.—*Vertigo from singing high notes.*

Miss H., a professional singer, was seen in January, 1879, on account of a suppuration of the right middle ear and a whistling tinnitus, which had begun two months before.

H. D., *R*. $\frac{5}{40}$, *L*. $\frac{40}{40}$.

Tuning-fork on teeth heard best in *R. E.* Right drumhead perforated posteriorly. Left showed a cicatrix (?) in front of malleus.

Vertigo was caused by singing a high note, and sometimes such notes sounded false to her. All notes of a piano were heard correctly.

CASE 5.—*Mastoid abscess without any evidences of disease of the external or middle ear.*

Wm. H., æt 6, came on September 28, 1876, complaining of pain in the region of the ear. There was redness, swelling and tenderness over the mastoid process. On September 30th an incision was made and a considerable quantity of pus evacuated. This was followed by recovery. No caries was detected. There was not the slightest evidence of any congestion or inflammation of the external auditory canal or middle ear.

This case is somewhat similar to those reported by Dr. D. Webster, in the ARCHIVES OF OTOLOGY, vol. viii, No. 1.

Such cases are rare.

CASE 6.—*Mental depression from impacted wax.*

Mr. T., æt 18, has been seen at intervals for several years on account of a chronic suppuration in the right middle ear. The left ear was normal. On May 15, 1879, the right ear was in very good condition ; the hearing was $\frac{2}{40}$, and there was no discharge. Patient came again on September 24th, complaining that since June he had suffered from "a feeling of heaviness in his head." Was "unable to concentrate his mind on anything for more than a few minutes." Felt as if he must give up his studies (in which he was very much interested), and wished to know whether he must leave college. Thought his deafness had increased, but had no pain, tinnitus or discharge. The patient was sullen and very despondent. Otherwise his health seemed to be excellent. He was very reticent by nature.

H. D., *R.* $\frac{C}{4\cdot5}$. External auditory canal filled with hard wax. After removing the wax, the hearing became $\frac{18}{40}$, and the tympanic cavity looked as it had at former visits ; there was no discharge. The patient obtained speedy relief, and in a few days reported the discomfort about his head gone. He was then as cheerful as usual.

This case was interesting, as illustrating the disturbing influence of impacted wax, even with an entire absence of tinnitus.

Mucus in the tympanum.

Within the past few months a number of cases of chronic suppuration of the middle ear have been seen, in which there were large accumulations of mucus in the tympanic cavity. It is not meant that there was a large admixture of mucus with an ordinary purulent discharge, but that the tympanum (and probably the mastoid cells) was filled with such masses of tenacious, glue-like material as are sometimes found with an imperforate drumhead. In some of the cases a recent purulent discharge seemed to have been replaced by the secretion of mucus ; in other cases there had been no discharge of pus for a long time. The symptoms were the familiar ones of oppression about the head, of feeling of *pressure*, of embarrassing fluctuations in hearing power, etc. The usual difficulty was found in removing the mucus thoroughly, and it re-formed in each case several successive times. This condition is not common in cases of chronic suppuration, in the reporter's experience. Other practitioners, however, may have seen it often. · Sea-bathing seemed to have a causative influence in two of the cases alluded to above.

An ordinary lachrymal syringe, with a long flexible nozzle, has been found very efficient for sucking out mucus from the drum, especially after a paracentesis.

NOTE ON THE TREATMENT OF ACUTE SUPPURATION OF THE MIDDLE EAR.

By Dr. EDWARD T. ELY, New York.

THE tendency to spontaneous recovery, manifested by so many acute diseases, is observable also in acute suppuration of the middle ear. Probably this is not a new thought to any reader of this paper, but it seems to the writer to be too much ignored in practice. Great labor has been required to lead physicians and laymen to consider acute suppuration of the middle ear as of any importance. This work has involved much writing and discussion as to the nature of the disease, and as to the necessity for prompt and efficient treatment of it. It is natural that many practitioners, having thus been laboriously awakened to its importance, should hold exaggerated ideas as to the remedies required for its cure.

Notwithstanding the efforts which have been made to bring patients with acute aural disease under treatment, the majority of them continue to be neglected by themselves and by their family physicians. The numerous cases of acute suppuration of the middle ear which have recovered, and which are constantly recovering, in spite of neglect or of bad treatment, afford proof of a tendency to self-limitation in this disease. Every aurist sees many patients who, in stating their history, refer to former suppurations of the drum which have ceased spontaneously. The drumhead is found to be well healed, although it may present extensive cicatrices, and the hearing is either perfect or only slightly

impaired.   It cannot be denied that many of these patients
have fared as well as if they had been under the most skil-
ful management.

Admitting these facts, should they not influence our practice somewhat?   It is not intended here to underrate the
importance of having every case of this disease under the
observation of a competent surgeon from the outset.   Nor
is it designed to make any argument against the greater
part of the treatment usually employed, but simply against
the use of astringents or caustics before they are certainly
indicated.   We are assuming that the pain and congestion of the first stage have been subdued, and that we
have to deal only with a perforated drumhead and a suppurating tympanic cavity.   Under these circumstances,
would it not be preferable, in every case, merely to keep
the ear clean and to watch it for a few days, to see what it
is disposed to do for itself, before resorting to any more
active treatment?   It will surprise a person who has never
done this, to find how often the drumhead will heal and the
disease be cured before this watching-process is finished.
The application of an astringent or caustic is certainly
needless in many instances.   The use of them, moreover,
has certain disadvantages.   If, in such a condition as we
are considering, the surgeon immediately applies them, he
complicates the problem before him.   If the ear does
not happen to do well, he is at a loss to know how
far this is due to the disease, how far to erroneous treatment.   Any person who has treated a severe case of purulent ophthalmia, threatening destruction of the eye, knows
how embarrassing our uncertainty as to the choice of remedies may become.   If, on the contrary, a suppurating tympanic cavity has been watched long enough to determine its
natural tendency, any needed remedy can be adapted to it
with far more accuracy.   The choice of even such mild
remedies as our weakest solutions of zinc or alum is not a
matter of indifference.   We have all seen cases where they
seemed to increase the swelling, or the discharge, or the loss
of tissue.   The following one seems to show a still more
serious effect:

Miss H., aged 20, consulted me November 30, 1877, with acute suppuration of her left middle ear of ten days' duration. There was a free discharge of pus, and no pain or swelling. I ordered syringing of the ear, and the instillation of a two grain solution of sulphate of zinc twice daily. Immediately after using the zinc drops she began to have violent pain in the ear. This pain continued all night, and, when I saw her the next day, the auditory canal was so swollen that the drum could not be seen ; the whole of that side of the face was swollen and tender, and there was congestion and pain in the eyeball. There was a temperature of 101° and some vertigo. Leeches, hot water, morphine, and rest in bed were prescribed. The pain, swelling and vertigo did not disappear until the evening of December 4th. I always attributed this attack to the effect of the zinc, although I have no further proof of the fact than the patient's own belief of it, and the history of the case.

The following cases are offered in illustration of what has been said above. Only a few are given out of a larger number which might have been presented, had it been thought essential to the argument :

I. Susie M., aged 6, came on November 11th with a history of pain in her left ear from six o'clock until eleven of the previous evening. The drumhead was found congested and ruptured, and there was a purulent discharge. Syringing of the ear with warm water twice a day was ordered. On the 14th there was no discharge, and the perforation seemed to be healing ; the syringing was discontinued. On the 16th the perforation had healed and the hearing was fully restored.

II. Miss J. H., aged 21, came on March 11th, having had severe pain in her left ear since 3 A. M. The drumhead was found ruptured, and there was purulent discharge. The hearing on that side was $\frac{6}{10}$. Leeches and the hot douche were ordered, and they seemed to arrest the pain at once. After that, the ear was simply syringed occasionally with warm water. On the 13th the perforaration was nearly closed. On the 18th it was completely healed, and the hearing was $\frac{40}{40}$.

III. Mrs. M.. aged 35, came on March 17th, saying that she had had a cold in her head for the past week ; that two or three days ago, while blowing her nose, she had felt a "cracking" in her right ear, and that since then there had been a discharge from the ear. Before this trouble the drumhead on that side was cicatricial from a suppuration in childhood. A large perforation was found in the posterior part of the drumhead, with a muco-purulent discharge. The hearing was $\frac{6}{40}$. Syringing with warm water, two or three times a day, was ordered. On March 19th the perforation was much smaller ; the discharge was still abundant. On March 20th there was no discharge. The next day her cold became worse, and she had some fever. The following three days she had throbbing and tinnitus in the right ear with reappearance of the discharge ; also had some vertigo. Was taking quinine during this time. On the 25th the discharge had ceased, and a few days later the perforation was healed. Hearing $\frac{6}{40}$.

IV. Mr. W., aged 40, came on February 24th with a broken drumhead and acute suppuration, in the right middle ear. The discharge had appeared on the 19th, after eight hours of pain in the ear. Syringing with warm water was prescribed. On February 27th, the discharge was found to be less. On March 2d, the discharge had ceased and the perforation was very small. A few days later the, drumhead was found to be healed and the hearing restored.

V. Master L., aged 5, came June 17th with a history of ear-aches, both sides, for the previous four weeks. An examination showed perforation of both drumheads and acute suppuration of the middle ears. No treatment was employed except syringing with warm water. The patient made a perfect recovery.

VI. Master F., aged 14, came on April 7th with acute suppuration of the left middle ear. The use of the warm douche was prescribed. On April 17th the ear was doing well, and the hearing was $\frac{18}{40}$. A few days after this the patient was cured.

In this case and the preceding one the exact date of recovery was, unfortunately, not recorded.

VII. Miss M, aged 18, came on December 14th with acute suppuration of the right middle ear, of a few days' duration. She had already had a chronic suppuration of that ear, following measles, which had been checked, without restoration of the drumhead. Warm syringing was prescribed. On January 14th the discharge was found to have ceased.

VIII. Master V., aged 16, came on June 20th with an acute suppuration of the left middle ear. The discharge, which was very bloody, had been noticed by the patient a day or two previously, after a night of very severe pain in the ear. There had already been marked deafness on both sides, from chronic catarrh, for many years. The only treatment prescribed was syringing of the ear with warm water two or three times a day. On June 27th the drumhead was found to be healed. There had been no discharge for several days.

The cases given above are thought to be sufficient in number and variety for the purposes of this paper. The local treatment in all consisted simply in syringing the ear with warm water as often as seemed advisable. Of course, the throat and the general health received attention when it seemed needed. It is believed by the writer that treatment as simple as this is sufficient for many cases of acute suppuration of the middle ear, and that it is usually well to make a trial of it, for a few days, before resorting to anything more energetic.

Several of the cases here presented are from the practice of Dr. D. B. St. John Roosa, to whom I am indebted for the use of them.

A CASE OF ACUTE INFLAMMATION OF THE MIDDLE EAR, WITH INFLAMMATION OF THE MUSCLES OF THE NECK, AND FACIAL PARALYSIS OF THE SAME SIDE. RECOVERY. WITH SOME REMARKS UPON THE INDICATIONS FOR WILDE'S INCISION AND TREPHINING THE MASTOID PROCESS.

By Dr. D. B. St. JOHN ROOSA, of New York.

THE following case gave me so much anxiety on account of a difference of opinion occurring between very competent authority and myself, as to the true significance of some of the serious symptoms, and as to the proper treatment to be pursued, that I report it, hoping it will be as instructive to my professional brethren as it has been to me.

May 5, 1879. Dr. S., æt. 45, a busy surgeon and medical journalist, consulted me in regard to uncomfortable and painful sensations in his right ear. He was somewhat anæmic, jaded from overwork, and he had an anxious appearance. He described the pain as extending from the right Eustachian tube to the drum, laying great stress upon the pain along the tube. The drumhead was red, the auditory canal normal. There was nothing marked about the pharynx. The hearing distance was not noted. Leeches were ordered to be applied to the tragus. I afterward learned that he had slight nasal catarrh and headache with pain in right lower jaw, on May 4. The next day I received a note from the patient stating that he did not feel able, on account of the pain, to come to my office, which was a very short distance from his. I found him in bed and apparently suffering very much. He complained of a pain like that from neuralgia, extending over the right side of the scalp, face, neck, the right auditory canal, and the Eustachian tube. Leeches and the hot douche were prescribed. The patient then told me that he had suffered very severely a few weeks before from facial neuralgia, that he then had no aural trouble, that he had had very lately an inflammation of the muscles of the opposite side of the neck. The membrana tympani was vascular but not bulging. Knowing that this patient had been very much overworked, with an insufficient quantity of fresh air, and seeing that he was pale and hyper-sensitive, I considered the pain as out of proportion to the objective symptoms of inflammation, and I therefore made a diagnosis of non-suppurative inflammation of the middle ear, with neuralgia of the fifth and seventh pair. In other words, I believed that the otalgic symptoms predominated over those of true inflammation. Warm applications behind and over the ear were advised. as well as the use of the hot douche. The hot douche was not well borne, nor was there much relief, except at short intervals, from these measures. It should also be said that I laid great stress upon maintaining the nutrition, and a generous diet was insisted upon. On the fourth or fifth day, the auditory canal was somewhat swelled but not tender. I incised the drumhead, but no pus or mucus was evacuated. The hot douche was now freely used and afforded relief. A very moderate suppuration occurred in the tympanic cavity. Morphia was administered, *pro re nata.* The patient sat sometimes out of bed, but did only tolerably well, complaining at intervals of very severe neuralgic pain which was relieved by morphia. He took nourishment badly except in the intervals of freedom from pain. He was very much depressed in spirits. There was no tenderness or any other inflammatory symptoms on the mastoid or in the pre-auricular region. On May 15th—ten days after I first saw the patient, I went out of town to fill a professional engagement, and my associate, Dr. E. T. Ely, took charge of the case until May 25th, and his notes are as follows :

" Dr. S. seems to be a case of acute suppuration of the middle ear, with considerable swelling of the auditory canal, slight discharge, no pain. 16th,—more pain and swelling, no discharge. 17th,—severe pain in whole right side of face and head and in the ear, not controlled by douche; no discharge; funnel-shaped swelling of the canal, not very tender. Consultation with Dr. A. H.

Buck. It was decided to incise the canal and re-open the drum-head. This was done under ether. The opening in the drum-head was very free, and the canal was incised from the bottom to the entrance. Three leeches were then applied to the tragus and one to the mastoid. Hot douche was continued. No pus followed these incisions.

May 18*th*.—Pain most of last night. A litt e easier this morning. Discharge of pus beginning.

May 19*th*.—Comfortable until evening, then great pain in ear and head, temperature 101½°; three leeches to mastoid, douche, morphia. 20th, not much pain ; weak and depressed. A. M. T. 98½°, P. 88, P. M. T. 100½°, P. 88. Slept most of the day.

May 22*d*.—No fever yesterday or to-day, one attack of severe pain last night, canal red and swollen, free discharge since incision: four leeches applied, and hot douche for twenty minutes every two hours.

May 24*th*.—Pain part of every day, no fever: severe pain last evening quieted by morphia; slight mastoid tenderness and œdema last evening and this morning, less swelling in canal. Dr. Buck was again called in consultation; he advised opening the mastoid by trephining. Dr. C. R. Agnew was called in the afternoon. He considered the case a typical one of mastoid disease of pro-liferous nature, but that no suppuration was going on there. He thought the disease was chiefly in the mastoid from the outset, and that there was meningeal congestion. By the ophthalmo-scope the veins in the right fundus seemed a little fuller to Dr. A. and to Dr. Ely than in left. Dr. A. advised potass. iodide gr. x. t. i. d., and increased to point of tolerance. Fl. ext. Ergot ℥ j. t. i. d., sodii bromid, gr. xv at night. Keep ear and mastoid warm with cotton, and omit douche."

May 25*th*.—Very slight œdema and some tenderness over mas-toid, and although only one dose of the iodide was taken, iodism was produced. Patient was awake all night from sneezing, and had some pain in the other ear. He is nervous and hysterical, buries his head in the bed clothes, and refuses to be comforted. He expresses the belief that he will not recover. On this date I met a gentleman with very large aural experience, and we went over the case very carefully. The patient seemed to be suffering very much and he located the seat of his pain by spreading out his hands like a fan over the right side of the head. The tender-ness about the ear was not very great, and was found in the neck

and occiput as well. The ear was discharging freely with healthy
pus. The mastoid was so slightly œdematous that I thought its
condition might be due to the leeches and other applications. It
did not seem to me to be a case of mastoid periostitis, nor did I
think there was any meningitis or cerebral disease. Although I
did not feel so sure of the former point as of the latter, I still
thought the pain was neuralgic rather than inflammatory. Inas-
much, however, as Dr. Agnew had on the day before given the
opinion that the mastoid was markedly involved, and that there
was a meningeal hyperaemia, and as the gentleman now in consultation was much more decided in the opinion that the mastoid
was the point of the origin of the pain, and, moreover, since my
own judgment was a little doubtful and wavering, I advised that
a Wilde's incision be made at once. If this incision failed to de-
tect disease of the bone, I resolved to take no further operative
steps at this time, although the gentleman in consultation afterwards stated to me, that he considered this but a step in the right
direction, he believing that the bone should be opened, and that
even if no pus were found, the bone-fistula would do no harm.
The incision was accordingly made ; no disease of the bone was
found. The wound was dressed to the bottom with lint, and a
poultice was applied.

May 28th.—The pains in the head and neck are not at all
relieved except when morphia is used in full doses. The tissues of
the mastoid, pre-auricular region and neck were red, swelled, and
tender at various points. These symptoms have increased since
the incision. The depression of spirits continues, but at times
the patient can be made quite cheerful by light conversation, and
after a dose of morphia. He is taking a moderate amount of
stimulants, and milk quite freely. Dr. William A. Hammond was
called in consultation : his opinion was that there was no disease
in the cranium, and that the pain was due to neuralgia largely
modified by malaria. He advised that 60 gr. of quinine be given
in twenty-four hours, for two days, and that this treatment be followed up by small doses of arsenic. This treatment was followed
by an apparent alteration of the pain, and not so much morphia
was needed.

On June 3d the muscles of the neck were so much swelled that
we pronounced them in a state of inflammation, and leeches were
applied. The arsenic and generous diet, as far as patient would
take it, with moderate doses of alcohol, were continued. The

neck was especially tender where nerves made their exit. There was no especial tenderness on the mastoid ; the patient could scarcely move his head from side to side.

June 7th.—The conjunctiva and outside of lids of right eye are reddened ; the ability to close the right eye is impaired.

June 8th.—Conjunctiva and lids less red than yesterday. Slight enlargement of gland at the angle of the jaw on right side. Severe pain in the jaw and mastoid region. Morphine was freely administered hypodermically for its relief. A poultice was kept on the side of the face and the head. T., 101½° : P., 100.

June 9th.—Swelling at the angle of the jaw increased ; pain severe, and facial paralysis on the right side well marked. The right lid does not completely close in winking. The right side of the face appears rounder and fuller than the left, and the mouth is slightly drawn toward the left. The tongue protrudes in a direct line, and there is no deviation in the uvula. There is apparently no disturbance of the sense of smell. T. 99½° ; P. 94. Two leeches were applied behind the ear.

P.M.—Severe pain ; mx. of Magendie's solution every three hours (hypodermically).

June 10th, A.M.—T., 98¼ ; P. 100. Slept well ; took about 1 qt. of milk during the night. Facial paralysis increased. Ophthalmoscopic examination by Dr. Roosa. The appearance of the fundus is the same in both eyes, and nothing abnormal is seen in either. The ear discharges freely.

P.M.—Longer intervals of freedom from pain. No morphine since the 8th at 9 P.M.

June 11th, A.M.—T., 99½°. Swelling at the angle of the jaw diminished. No pain since June 10th at 9 P.M.

P.M.—Pain recurs ; not so severe. Chloral and bromide of sodium are given for its relief.

June 12th, A.M.—Patient slept badly. Pain returned in the old regions, the jaw, behind the ear, and over the right side of the head. T., 98¾° ; P. 94. Patient very much depressed in spirits. Morphia again administered. At 5 P.M. a consultation was held, at which were present Dr. Alfred L. Loomis, Dr. Henry B. Sands, Dr. Charles R. Briddon, Dr. W. M. Carpenter, and the attending physician, Dr. Roosa. After Dr. Roosa's statement that the pus was freely discharging from the auditory canal, and that, in his opinion, there was no retained pus in the bone, without claiming to decide the strictly *aural* points of the case, positively, the conclusion was reached by the consulting surgeons and physicians that the patient had no symptoms of intra-cranial trouble, that there was no indication for operative interference with reference to the mastoid process, or suppuration in any part of the neck ; that supporting treatment was demanded. On the suggestion of Dr. Loomis the stimulant he was receiving was increased to 1½ oz. of whiskey every three hours, and pushed to 2 oz. as soon as it became evident that it did not disagree with his stomach.

June 13th.—Patient feels very comfortable ; has slept well, is taking 2 oz. of whiskey in a tumbler of milk every three hours, and has not experienced the slightest intoxicating effect. Takes nourishment aside from the milk. T., 99° in the morning, 98¾° 6 P.M. ; pulse, between 96 and 100. Patient also takes citrate of iron and quinine. At 8 P.M. patient again complains of severe pain. Morphia administered at 9.30 P.M. At 3 A.M. on June 14th he was seen by Dr. Ely on account of great pain. Morphia was given at that time and one hour later. At 8 o'clock the pain was still unrelieved, and the swelling about the angle of the jaw and the mastoid process was very much increased. Morphia was freely administered *p. r. n.*, and a consultation was held at 1.30 P.M., at which three aural surgeons and one general surgeon were present. The following opinions were given : Dr. ⸺, an otologist, saw no indication for operative procedure, while he believed there was mastoid disease. Dr. ⸺, also an otologist, believed that the patient was suffering from mastoid disease, and that trephining should be performed at once. Dr. ⸺, aural surgeon, thought there was no serious internal trouble, that it was external, and that the patient was probably suffering from some kind of poisoning—malarial ? sewer gas ? that no operation was advisable. The general surgeon thought that pus would be found somewhere about the stylo-mastoid process, and he thought that nature would relieve the patient by suppuration. He laid great stress on the continued application of poultices, and he was not in favor of operative interference to-day. Dr. Roosa adhered to his original opinion, that the patient had a moderate inflammation of the middle ear, with great neuralgic pain, and that the swelling of the neck and facial paralysis may have been caused by the operative procedures already undertaken, and that trephining was not justifiable, but that it would be injurious. It was decided to continue the alcohol and to make the application of poultices very thoroughly over the neck and mastoid.

An examination of the urine on June 15th gave the following result : Dark straw-color, acid, sp. gr. 1024, albumen in moderate quantity, casts 2, slightly granular, uric acid a little, pus a little, mucus a fair amount, oxalate of lime a little. June 15th, the ear is suppurating moderately. The drumhead is granular, canal moderately swelled, ear easily inflated by Politzer's method. The swelling in the course of the sterno-cleido-mastoid muscle, and about the neck, seems to be increased, but the tenderness is not so marked. The symptoms point to abscess forming in the connective tissue, and in the muscles of the neck, and over the mastoid process, Dr. Roosa does not think there is retained pus anywhere in the head, or inside of the temporal bone. There is a particularly tender point, 1¼ in. in a direction directly backward and a little downward from the lobe of the ear. There is scarcely any œdema about the Wilde's incision. T., 99°. P,, 100.

3 P.M.—The swelling has begun to subside. Dr. ——, a general surgeon who had seen the patient on the 13th, saw the patient this afternoon, and thinks it possible there is pus in the petrous portion of the temporal bone, and that the swelling may be due to a temporary plugging up of the communication with the tympanic cavity.

Dr. Roosa thinks there may be pus in the cellular tissue, but does not think that it is necessarily connected with the tympanic cavity. The treatment was continued.

June 16th.—P. 98. T. 99°. Patient slept well. Dr. Roosa opened the track of the Wilde's incision with a probe. The swelling and œdema in the mastoid process and about the angle of the jaw remained the same.

Another consultation was held during the day, at which there were present two general surgeons, two otologists, and Drs. Roosa and Carpenter. One of the surgeons expressed the opinion that the patient's general condition had improved since he last saw him, but he declined to express any opinion in regard to the necessity for operative interference with the ear. He believed it *possible* that the operations already performed might have aggravated the symptoms. The other general surgeon inclined toward trephining the mastoid. This should certainly be done in his opinion if there is a probability that there is not a free opening from the mastoid cells into the tympanic cavity, and this was a point to be decided by the aural surgeons. One of the otologists thought the patient better, and that no operation should be done. The other aural expert believed that the bone should be opened. Dr. Roosa stated that his opinion was unchanged, but that he had so much respect for the opinion of the gentleman who was so decided with regard to the necessity for an operation, as well as for that of the one who was inclined towards it, that he wished for further advice before he declined to open the mastoid. By agreement Dr. Robert F. Weir, who was for some years aural surgeon to the Eye and Ear Infirmary, and who is now surgeon to two general hospitals, was invited to see the patient independently and alone, at 9 o'clock this evening, without knowing any of the opinions that had been expressed, until his own was formed. Dr. Weir gave the following opinion : that the disease is probably an inflammation extending down the external auditory canal, in the angle close to the point where the facial nerve passes, and that it may perhaps involve the mastoid process : he is inclined to think it does not : there is no indication for surgical interference for the present. The general plan of treatment was therefore continued. June 17th. An examination of the urine made this day shows specific gravity 1020, and a well-marked trace of albumen. No casts. The general condition of the patient is improving, and the swelling about the neck is subsiding.

June 19th.—Patient is still doing well. Treatment has been continued.

June 21st.—Patient sits up and walks about, swelling of the neck nearly gone, no pain or tenderness, drumhead healed, hears the watch tick. The swelling and redness of the neck reappeared for one day, while the patient was convalescent, and alarmed him, but it passed away in a few hours. This was after the drumhead had healed. July 5th. Patient walks about, has been twice to the sea-shore, hears the watch one inch, facial paralysis improving under electricity. July 20th. A note from the patient states that he can hear the watch ten inches, the voice as well as ever, that his facial paralysis is gone, that he considers himself well.

Remarks.—I regret very much that the early notes of this case are not more full; yet I think they are sufficiently so to give my readers a fair idea of the first symptoms. It is probable, however, that the mere recital has not conveyed to the minds of those who have followed it a full sense of its doubtful features. They were such that, taken in connection with the patient's high professional position, they gave me great anxiety lest I should omit to do my full surgical duty to the case. The more recent of the notes were taken stenographically by Dr. W. M. Carpenter, to whom the patient and I are indebted for intelligent and assiduous care.

The point to be settled during the course of the disease was this : Is there a hidden suppurative process going on in any part of the temporal bone which causes the pain, œdema, tenderness, cellulitis, myositis and paralysis of the facial? My answer to the question was, No. The severe paroxysmal pain did not arouse the suspicion in my mind that there was mastoid disease, because there was absolutely no well-defined tenderness, redness or œdema until leeches and poultices had been freely applied, and not until two paracenteses of the drumhead and very free incisions of the auditory canal had been made.

On the 25th day of May, when I saw the patient after an absence of ten days, there was certainly a moderate amount of œdema, and this led me, although I suspected it had been caused by the leeching, to advocate a Wilde's incision, especially as I then thought it a harmless procedure, and two otologists, who had seen the patient with Dr. Ely, thought the disease markedly involved the mastoid, although only one of them advocated any operative procedure. I now think that this incision was a mistake, and that to it we owe the increase of the inflammatory symptoms in the neck and the facial paralysis. Indeed I now believe, on a calm looking over of the case, that every operative interference, from my first paracentesis down to the Wilde's incision, was unnecessary, and that the traumatism needlessly aggravated the painful case. The key-note was struck in the proper management of the case, in my opinion, when the supporting, anodyne and anti-malarial treatment by means of milk, alcohol, morphia and quinine was vigorously entered upon.

I believe, furthermore, that the disease would have been more easily subdued if I had gotten the patient out of his house and by the sea-side, before the graver symptoms set in. This I urged upon the patient and his friends, but without avail. It was simply a case of sub-acute, non-suppurative inflammation of the Eustachian tube and tympanic cavity, occurring in an anæmic and, consequently, neuralgic and hysterical subject. That he was anæmic was not only noted by me at my first interview, but when Dr. Loomis was called in consultation he stated that he had noticed the doctor's anæmic condition for a year.

Neuralgic he certainly was, for he had barely gotten through with a severe attack of facial neuralgia when the trouble occurred in the ear. The character of the pain during the whole course of the disease was not that arising from deep-seated trouble in the middle ear, but rather of a disease like neuralgia, in which there is an intensity at different times, and which has intervals of complete cessation. It was sometimes easy to divert the patient by light conversation or an anecdote, for quite a long time, and on some few occasions the use of water in the hypodermic syringe was followed by as much effect as the employment of morphia. Now, the character of a pain caused by severe inflammatory action in the tympanic cavity or mastoid process is such that no physician who has seen much of it would attempt to alleviate it by any diversion of the patient's spirits or by a placebo. Only positive means, such as local bloodletting or division of the periosteum, will subdue this. I have long since recorded my experience * that morphia alone will not mask the severe pain of an acute inflammation of the middle ear. As Von Tröltsch aptly says, an inflammation of the tympanic cavity is essentially a periostitis, and every surgeon knows of what little avail are drugs against the pain of this disease, except when it occurs as a result of the deposition of syphilitic poison. It should have been said before that this patient had no syphilitic taint whatever.

I considered the patient to be nervous and hysterical, be-

---

* Transactions of the American Otological Society, 1875, page 89.

cause he bore his pain very badly, and because he suffered from very great depression of spirits. It is not usual, in my experience, for a patient suffering from acute inflammation of the middle ear, to dwell very much on his prospects of recovery, or to be greatly depressed about his future. He is generally taken up so much with the severity of his pain as to have room for nothing else. Then there was something in the history of the house in which the patient lived, which I failed to impress upon some of the gentlemen who saw him with me, which led me to believe, as was once independently suggested by Dr. Noyes, who saw him two or three times, that there was an element of blood-poisoning in the case, perhaps from sewer gas. Two members of the family had suffered from acute aural disease a few months before, and an examination made by competent authority late in the course of the case, showed that there was an escape of sewer gas in the cellar. I do not know that any special significance is to be attached to the presence of albumen in the urine, but so far as it goes, it indicates a somewhat deteriorated general condition. In analysing the case, I come over and over again to the conviction that the operations did harm. That traumatism such as the patient experienced in the paracentesis, and in the very free subsequent division of the membrana tympani, and the free incisions in the auditory canal, and the cut down to the mastoid bone, might induce adenitis, myositis, cellulitis, and that facial paralysis might result from pressure upon the nerve as it makes its way out of the stylo-mastoid foramen, I think does not admit of a doubt. Certainly there never was any evidence that the facial suffered any lesion until after it had left the cranium and tympanic cavity. Besides, the swelling and paralysis occurred at a point of time which makes it possible to believe that traumatism may have caused them. But, the crucial test of the correct diagnosis was in the results of the case. There was no escape of retained pus either from the mastoid or from the neck. It certainly was not pus which caused the serious symptoms. When they were at their height the discharge from the ear went on, but gradually diminished. And when the patient was fairly convalescent, and up and about, the old swelling and redness of the neck reappeared for several hours. Besides, it should be noted that no chill occurred during the progress of the case. This fact, together with the clearness of the patient's intellect, gave me great encouragement when I was struggling against the opinion of a valued colleague who thought the patient was dying for want of an operation. Dr. S. was relieved after large doses of quinine at a time when the pain was intense, and when these seemed to fail, he was permanently cured after the full doses of alcohol advised by Dr. Loomis.

I believe that I was the first in this country to formulate the symptoms which should lead to the prompt performance of Wilde's incision, and trephining the mastoid process. As bearing upon this discussion, I venture to reproduce these formulæ here.

* I.—The integument and periosteum should be freely divided over the mastoid in all cases in which there is pain, tenderness and swelling in the part.

II.—Such an incision should also be made whenever severe pain, referred to the middle ear, exists, and is not relieved by the usual means, *i,e.*, leeches, warm water, etc.

III.—An explorative incision should be made when we have good reason to suspect the existence of caries and retained pus in this part.

IV.—The mastoid bone should be perforated after such an incision whenever the bone is found diseased, or a small fistulous opening should be enlarged. It should also be perforated when we have good reason to believe that there is pus in the middle ear or mastoid cells which cannot find an exit by the external auditory canal.

I omit the fifth rule, as it has no bearing upon this case.

As is well known, it is very difficult to formulate rules for operation which shall cover all cases. All rules must yield to peculiar circumstances. Still, I think my first formula might be a little more guarded. I would now write, " the integument and periosteum should be freely

---

* Treatise on the ear.—Roosa.   Page 424.

divided over the mastoid, where there is pain, *chiefly re-ferred* to this region, as well as tenderness and swelling." When the Wilde's incision was made in this case, upon my advice, the pain was not "chiefly referred" to this part. But there was no particular tenderness, simply a very slight œdema which might have been due to the applications that had been made, so that, according to the rule, without altering it, the incision need hardly have been made in this case. The second formula, however, justified the incision fairly, and this, I think, should be modified. I would now write, instead of the second, "such an incision should usually be made whenever severe pain referred to the middle ear *constantly* exists, which is not even temporarily relieved by the use of leeches, warm water, morphia, quinine, etc." In the case we have been studying, the pain was referred to many parts besides the middle ear, and it was relieved for hours at a time by morphia, quinia and alcohol. I would not modify the third and fourth formulæ, for I still think that the bone " should be perforated when we have good reason to think that there is pus in the middle ear or mastoid cells, which cannot find an exit by the external auditory canal." Everything turns upon the "good reason to believe," and I did not advocate, indeed I could not consent to opening the mastoid, because I did not think that we had " good reason to believe " that pus was retained in this part. I am in full accord with the great English surgeon, Sir James Paget, who, in his admirable lectures, expresses many times his hesitation to perform any surgical operation, however trivial, that is not absolutely required. We have no right, I think, to perform operations to clear up doubtful diagnoses, if, in case the operation proves to have been unnecessary, the patient will be decidedly the worse for it. If we put ourselves in the place of our patients, what we may regard as a trifling thing, "a mere cut," will not be so esteemed. A mere cut, when unnecessary, may have the most serious consequences, and all the history and symptoms should be carefully weighed before even that is undertaken. Such care will never prevent prompt, rapid and thorough surgical interference when demanded.

In teaching medical students, I have always found them, when fully awakened to the dangers of *neglecting* certain diseases, to be more apt to do too much than too little, especially with the knife and active drugs. It is possible also that the crying ignorance and neglect of the previous decades in regard to the treatment of aural disease has had a tendency to cause us who see many of the affections of the ear, to lean toward the side of surgical operations upon the drumhead and mastoid, a leaning no less dangerous to the cure of some cases, than was the steering toward Scylla or Charybdis to the safe navigation of ancient mariners.

CLINICAL CONTRIBUTIONS.*

By D. B. St. JOHN ROOSA M. D., and EDWARD T. ELY, M. D.

(With a wood-engraving.)

I.

AMBLYOPIA FROM QUININE (?)

A CASE of amblyopia from supposed cinchona poisoning was published by Dr. Roosa in these archives, vol. viii., No. 3. Dr. Miranda, of this city, after reading the account of this case, informed us that he had met with a similar one, and was kind enough to induce the patient to pre sent herself for examination. Through his courtesy, the following notes were obtained:

Mrs. B., aged 34, was seen on Nov. 7, 1879. She had a pernicious malarial fever in Cuba, in Nov., 1877. Upon one day she took 50 grains of the sulphate of quinine, 90 grains the next day, and 30 to 40 grains for the succeeding three or four days. At 11 P.M. of the day on which she took 90 grains of quinine she became blind, and remained so for the next three or four days. After that a gradual restoration of sight occurred. There were no aural symptoms. The intellect was clear most of the time. There was occasional delirium when the fever began, and when she was taking the large doses of quinine. The blindness was so complete that she had no perception of light. The testimony of her family physician and of her friends agrees with her own upon this point. No ophthalmoscopic examination was made. .

She says that her vision was perfect before her illness, and that she has never been as well since. Feels as if there was a veil over

---

* This paper was received Nov. 2, '79.—EDITOR.

her eyes. Cannot tell whether her linen is clean when it comes from the wash. Cannot do "shopping," because she no longer distinguishes colors well. Can see certain shades of dark-blue well, but distinguishes all other colors very imperfectly. Has most trouble with the different shades of red. When she first began to recover from blindness, had no color-perception whatever.

R. E., V.=$\frac{20}{40}$. Reads 1 Jaeger fluently.

L. E., V.=$\frac{20}{40}$. Reads 2 Jaeger with difficulty.

V. =$\frac{20}{30}$—with both eyes open.

With the ophthalmoscope, the fundus of each eye looks somewhat indistinct ; the discs look too white ; the capillaries seem deficient.

The visual fields (tested at 12″ by Carmalt's perimeter) are found concentrically limited, as is shown in the accompanying diagram, and are almost the same in both eyes.

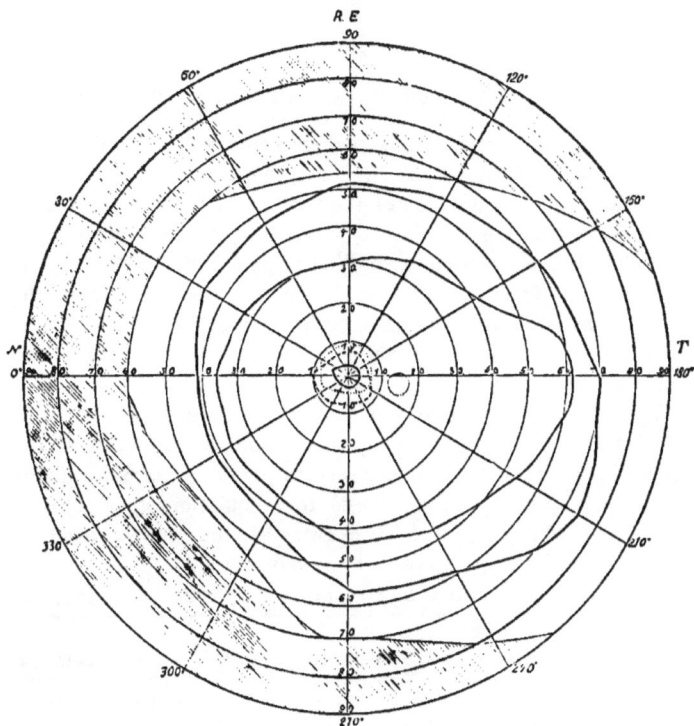

The color-fields of the left eye are also shown. The right eye was not tested so fully, on account of the fatigue of the patient. There were no scotomata. In the diagram the second line from without inward represents the field for blue; then follow consecutively the fields for yellow, green and red.

There is, of course, no positive proof that quinine produced the amblyopia which has just been described. It may have come from malarial poisoning, from a neuritis in the course of the fever, or from some other cause.

The case is offered, like the one before reported, merely as a slight contribution to the subject.

Von Graefe reported some striking cases of amblyopia which he thought must have been caused by the use of quinine.*

Sir James Paget, in a recent number of the Contemporary Review, says. "large quantities of quinine will make a man, at least for a time, deaf and blind."

II.

TOTAL AMAUROSIS FROM CONCUSSION. RECOVERY.

P. M., aged 25, iron-railing maker, came on September 23d, with the following history : On September 21st, while drunk, he was kicked on the right side of the face. When he became sober a few hours later, he found that he had lost the sight of his right eye. There was no redness, swelling or pain in the eyeball or lids at any time. Is sure the vision of that eye was good before the accident, as he used the eye constantly in *sighting* railings. A leech had been applied to the upper lid by a friend as soon as the loss of sight was discovered, and the bite "bled for an hour."

An examination showed some superficial contusions about the cheek and temple. No swelling except from the leech-bite on the upper lid. All parts of the eye seemed perfectly normal. There was no perception of light. Tests for simulation gave negative results. The vision of the left eye was $\frac{2}{3}\frac{6}{6}$, and not improved by any glass. The ophthalmoscope showed nothing.

Hypodermic injections of strychnia were ordered, chiefly for

---

* *Arch. für Ophth.*, Bd. 3, 2, p. 376

NOTE.—In a letter just received from Dr. L. M. Yale, he says that the patient whose case was reported in these ARCHIVES, vol. viii. No. 3, took a stronger tincture of cinchona than was at first supposed. Dr. Yale now estimates the amount of the alkaloid taken at 400–500 grains, instead of 125 grains as stated in the published report.

the purpose of keeping the patient under observation.   An injection of $\frac{1}{80}$ gr. was given at once.

September 24th.—Injected $\frac{1}{40}$ gr.   V. counts fingers at eighteen inches.

September 26th.—Injected $\frac{1}{30}$ gr.   V. $= \frac{8}{70}$.

October 24th.—Has not been seen since September 26th.   Says that he has been at work since that date, and that his sight has steadily improved ; but he is sure that it is not as good as before the accident.   Describes a slight *blur* over every object at which he looks.   Has not the slightest pain or discomfort about eye.

R. E., V.$=\frac{20}{60}$.   With $-\frac{1}{30}$c. 135°, V.$=\frac{20}{40}$.   L. E.   V.$=\frac{20}{40}$.   Not improved by glasses.   Reads 1 Jaeger at 6″ with either eye.   Right visual field concentrically contracted.   Measures 105° laterally and 80 vertically.   No scotoma.   With the ophthalmoscope the right disk appears much whiter than the left one (especially at the temporal side), and much whiter than on September 23d.   The vessels of the retina seem somewhat reduced in size.

Patient has not been seen again.   The recovery of sight in this case was not attributed to the strychnia.

OPHTHALMOSCOPIC OBSERVATIONS UPON THE REFRACTION OF THE EYES OF NEWLY-BORN CHILDREN.*

By EDWARD T. ELY, M.D.

DURING the past year I have made some ophthalmoscopic examinations to ascertain the refraction of the eyes of newly-born children, and I will venture to publish the results of my work, meagre as they are.

I have examined 111 children, of whom only six were over two months old. Among this number I have been able to make a satisfactory examination of 154 eyes in 90 subjects. Seventy-nine children were examined under the following conditions;

(1.) The eyes were put under the influence of a single instillation of a solution of atropine, containing two grains to the ounce, and a small dose of paregoric was given for its quieting effect.

(2.) The accommodation of my own eye was completely paralyzed by atropine.

(3.) The child was held erect by an assistant, and its lids were gently held apart, when necessary, by the thumb and forefinger of one of my hands.

(4.) The examination was made in the ordinary dark room, with moderate illumination from artificial light, by a Loring ophthalmoscope and in the upright image.

Following this method, I recorded the refraction of 105 eyes. For the sake of convenience, only 100 of these have been used in tabulating the results.

---

* This paper was received November 2, '79.—*Editor*.

I.

SEX.	EYES.	REFRACTION:		
		EMMETROPIC.	MYOPIC.	HYPERMETROPIC.
Males,	48	9	6	33
Females,	52	8	5	39
Total,	100	17	11	72

II.

AGE.	EYES.	EMMETRIC.	MYOPIC.	HYPERMETROPIC.
One week or less, .	51	11	4	36
1–2 weeks, . .	28	2	7	19
2–3 " . .	9	2		7
3–4 " . .	4	2		2
4–5 " . .	3			3
5–6 " . .	1			1
6–7 " . .	2			2
8 " . .	2			2
Total, . .	100	17	11	72

The nationality of the mothers could be obtained in thirty-nine cases;—

III.

CHILDREN.	NATIONALITY OF MOTHERS.	EYES.	EMMETROPIC.	MYOPIC.	HYPERMETROPIC.
23	Irish . .	38	6	5	37
9	American .	15	1	4	10
3	German .	5		2	3
2	English .	4			4
1	Bohemian .	1			1
1	African .	1			1
39		64	7	11	46

The refraction of the mothers' eyes could be obtained in ten cases only.

IV.

	MOTHERS.		CHILDREN.	
	RIGHT EYE.	LEFT EYE.	RIGHT EYE.	LEFT EYE.
I	H	H	H	H
2	H	H	{ H { H	H } H } Twins.
3	H	H	H	H
4	M	M	E	E
5	H	H	H	H
6	H	H	M	M
7	E	E	H	H
8	M¼	Not seen on account of syncope.	H	Not seen on account of conjunctivitis.
9	M	M	H	
10	H	H	H	H

Of the hundred eyes, as will be seen from Table I, 17 per cent. were emmetropic, 11 per cent. myopic, and 72 per cent. hypermetropic. A calculation based upon the number of children gives about the same result.

Forty-nine eyes in twenty-seven subjects were examined in a somewhat different way. The pupils were dilated by a single drop of a very weak solution of atropine, containing one grain to the pint. No paregoric was given. No atropine was used in my own eyes. Conditions (3) and (4) were the same as before.

V.

SEX.	EYES.	REFRACTION.		
		EMMETROPIC.	MYOPIC.	HYPERMETROPIC.
Males,	19	2	9	8
Females.	30	2	7	21
Total.	49	4	16	29

VI.

AGE.	EYES.	EMMETROPIC.	MYOPIC.	HYPERMETROPIC.
1 week or less, .	13		3	10
1–2 weeks, . .	20	2	7	11
2–3 " . .	8	2	2	4
3–4 " . .	2		2	
4–5 " . .	1		1	
5–6 " . .	4			4
7–8 " . .	1		1	
Total, .	49	4	16	29

VII.

CHILDREN	NATIONALITY OF MOTHERS.	EYES.	EMMETROPIC.	MYOPIC.	HYPERMETROPIC.
10	Irish . .	17		6	11
7	American .	13		7	6
3	German .	6			6
2	English .	4			4
2	Scotch . .	4		2	2
2	Swede . .	4	2		2
1	African .	1		1	
27		49	2	16	31

The following table includes all cases observed.

VIII.

SEX.	EYES.	EMMETROPIC.	MYOPIC.	HYPERMETROPIC.
Males,	72	11	15	46
Females,	82	10	12	60
Total.	154	21	27	106

Emmetropic about 14 per cent.
Myopic   "   18   "
Hypermetropic " 69   "

The appearances of the fundus in these eyes were similar to those seen in adults. There were the same deposits of pigment, and the same choroidal crescents and rings about the edge of the optic disc; there were variations in the

color and in the shape of the disc, and in the form and
position of its physiological excavation; and there were
irregularities of choroidal pigmentation. A difference of
refraction between the two eyes was not noticed in any case.
I believe these observations agree with those of Von Jaeger,
mentioned below.

As a whole, the disc and fundus were much lighter than
in adults. The crescents about the disc were found indis-
criminately in Emmetropia, Myopia and Hypermetropia.
The appearances of the fundus in the myopic eyes could
not be seen to differ from those of the others.

In case No. 77 the optic disc of the right eye presented a
large excavation, of grayish hue, simulating the appearance
of advanced atrophy in the adult. The margin of the excavation, at one part, appeared very abrupt, as in glaucoma.
The disc of the other eye looked normal.

In case No. 63 there was marked neuro-retinitis in the
right eye. The history of this case was as follows: The
child was left upon the steps of the Foundling Asylum one
cold night in January. Having been left at the wrong
door, it was not discovered until two o'clock in the morning.
It was then so chilled that it was revived only with difficulty. It had not been washed and was evidently only a
few hours old. It was seen by me in the afternoon of the
same day. It was a small baby, of " senile " appearance;
otherwise, the general examination showed nothing abnormal. The ophthalmoscope revealed marked evidences of
neuro-retinitis in the right eye. The optic disc and sur-
rounding retina were much swollen; the retinal veins
enlarged and tortuous. Scattered over the fundus were
some small hemorrhages, and a few round, yellowish spots,
looking like exudations. There was slight divergence of
the optic axes.

The next day, January 28th, the appearance of the right
eye was about the same. The fundus of the left eye looked
cloudy, as if the same process were begining there. The
eyes were examined by Dr. Loring and Dr. Roosa, who
agreed with my diagnosis.

January 29th, there was no special change in the appearance of the eyes. The baby was nursing well and looked better.

February 18th; the eyes seemed about normal. There was still slight divergence. The general condition was much improved.

Many efforts were made to obtain a specimen of this child's urine for analysis, but without success. Eleven months later, I asked permission to examine the condition of his eyes again, and was told that he had died of gastro-intestinal catarrh during the hot weather. I know of no other case on record where neuro-retinitis has been observed in so young a child.

Although somewhat irrelevant to the subject it may be stated here that the iris in every baby seen by me, including one negro, was of a bluish color. Dr. V. G. Culpepper has been kind enough to make more extended observations for me upon this point, in the Maternity Service of Charity Hospital. In fifty newly-born children noted by him, there was only one dark iris. This occurred in a negro. One of the nurses at the Maternity Hospital assured me that her attention had often been directed to this subject, and that in over 1,000 confinements she had seen only one child with dark irides. The mother of this child, she said, had very dark hair, eyes and skin. I think that all the text-books state that the eyes of all newly-born children are blue. This is undoubtedly the rule, but it does not seem to be invariable.

Many of my cases were very young, and it may be of some interest to consider these apart from the others. Forty-four children (yielding 64 eyes) were seen during the first week of life:

One child, a female, was examined 30 *minutes* after birth. The refraction of both eyes was hypermetropic; there was a black line of pigment about the discs; all parts of the eyes look well developed. The mother was hypermetropic in both eyes.

One male, seen 2½ hours after birth, was hypermetropic in both eyes.

One male, seen 5½ hours after birth, was markedly hypermetropic in both eyes.

One female, seen 6 hours after birth. Both eyes hypermetropic; optic discs very white; retinal vessels larger than usual.

Two males were seen 20 hours after birth. One had a hypermetropia of $\frac{1}{12}$ in each eye. The other was hypermetropic in the right eye; the left was not seen satisfactorily.

IX.

CHILDREN	EXAMINED ON	EYES.	EMMETROPIC.		MYOPIC.		HYPERMETROPIC	
			Under the strong solution of atropine.	Under the weak solution.	Under the strong solution.	Under the weak solution.	Under the strong solution.	Under the weak solution.
9	1st day.	14	1		1		10	2
5	2d day.	6	2				4	
3	3d day.	4	2			1	1	
8	4th day.	8	2				4	2
7	5th day.	8	2		1		3	2
2	6th day.	4				2		2
10	7th day.	20	2		2		14	2
Total, 44		64	11		4	3	36	10

According to Table IX, all three of the principal refractive states of the eye exist on the first day of life.

Certain details regarding the examinations may be interesting to some readers. Aside from the trouble of obtaining the subjects, the difficulties were found to be considerable. Most of the infants were very restless and disposed to cry. Even when the child remained quiet enough, there was often constant oscillation or rolling upward of the eyeballs when the light was thrown upon them. In some cases it was easy to get a general view of the fundus where it was impossible to get such a succession of views of some particular object as is required for measuring refraction. Or, the refraction could be ascertained in a general way, but the child would begin to cry, or become very restless, before an exact measurement could be made. This happened so often that the degrees of hypermetropia and myopia have been omitted from the tables altogether, even where they were determined. I should not think it possible to determine a myopia or hypermetropia of less than $\frac{1}{48}$ in these cases. I freely admit that I did not feel competent to do so.

Sometimes, success was prevented by mucus on the cornea, from catarrhal inflammations of the conjunctiva; in several cases, by an opacity of the cornea, caused by a very rapid drying of its epithelium as soon as the lids had been held apart for a few seconds. Any working back or forth of the lids to moisten the cornea, or to wipe mucus from it, was very apt to make the child cry. Anæsthesia from chloroform-inhalations was tried, but it was followed by such a rolling upward of the eyeballs as to prevent examination. A small speculum to hold the lids apart, and a fixation-forceps to steady the eyeball, were also tried. These instruments produced so much crying, congestion of the eye, and spasmodic pressure from the lids, that I abandoned them. After trying many expedients, the best way that I found for keeping the child quiet was to give it a small dose of paregoric, as already mentioned, and to allow it to suck the assistant's finger, or a bit of rag, moistened with syrup. It is much easier to hold the child quiet when a towel is wrapped around its body so as to confine the legs and arms. When it was necessary to hold the lids apart, my fingers served better than anything else. When the eye rolled out of view, a little pressure against it through the lid, with my finger, would generally cause it to move into a more favorable position again. The eyes seemed less likely to roll upward when the child's head was held perfectly erect. In using no instrument for fixing the eyeball, I found it almost impossible to see the fundus well through the undilated pupil; a solution of atropine of one grain to the pint always caused sufficient enlargement. Nothing that caused pressure upon the globe was used in any case here recorded.

The children were obtained in the Maternity service of Charity Hospital, the N. Y. Foundling Asylum and the

Marion Street Lying-In Asylum. Probably they were all from the lower classes of society. It was never possible to examine more than one parent, and that the mother. I did not even succeed in examining enough mothers to be of any value. At one visit the mothers would be to ill to be examined; and at the next visit, perhaps, they would be gone from the hospital. It was rarely possible to examine a child a second time when desired. The prejudice of the mothers against having their infants utilized in this way was always an obstacle. These facts will be readily understood by all who are familiar with the population of our public institutions.

My examinations were discontinued sooner than I had intended, owing to their interference with other work and to the personal discomfort caused by the constant use of atropine in my own eyes. I hope to be able to add to them at some future time. Some of the mothers of the infants examined by me have been followed to their homes and I may be allowed to keep their children under observation, and to examine their eyes from time to time, as they grow up, Doubtless, a series of observations of that kind would be of value.

The only similar investigations with which I am acquainted are those reported by Professor von Jaeger of Vienna.* He published the results of examinations of 100 eyes in fifty children between nine and sixteen days old. He found 5 per cent. emmetropic, 78 per cent. myopic, and 17 per cent. hypermetropic. The following table[1], copied from his work, exhibits his results:

Strength of correcting glass in Vienna inches.	+15	+25	+40	0	−60	−45	−30	−20	−15	−12	−10	−8	−6
Newly-born children, 9–16 days old.	9	8	5	1			3	26	20	14	10	4	

Von Jaeger concludes that the adjustment of the eye during the first days of life is predominantly a myopic one.[2]

---

* Ueber die Einstellungen des dioptrischen Apparates im menschlichen Auge. Wien., 1861.

1. *Id.* p. 20.　　　2. p. 10.

He says that this adjustment is due chiefly to a stronger curvature of the lens with a shorter distance of its anterior surface from the cornea.[3] He also expresses his belief in a congenital lengthening of the antero-posterior axis of the eyeball, causing a myopic adjustment,[4] and says that it is often easy to recognize in the eyes of the newly-born, the primitive forms of a conus.[5] He was led to the same conclusions, also, by measurements which he made of this same class of eyes in the cadaver. He affirms that the structure and adjustment of the eyes of the newly-born remain unaltered only for a short time—a few weeks—after which they change with the general development of the body.[6] " During the period of development, both hypermetropic and myopic eyes take on the normal structure ; and eyes which were originally adjusted for parallel rays are, after this period, adapted, rarely indeed, for convergent rays, but often for divergent ones."[7]

" The eyes of the children in the first years of life show, also in individual cases, a difference in their dioptric adjustment. The majority are, however, adapted for greater distances, when in a state of full relaxation of the accommodation."

" As, in general, the adjustment for smaller distances is characteristic of the eye of the newly-born, so is the adjustment for greater distances characteristic of the eye of the child proper."[8]

No attempt is here made to explain the great difference between von Jaeger's statistics and mine, especially with regard to the percentage of myopia. Our methods of making the examinations were somewhat different, as I have

---

3. p. 11.   4. p. 26.   5. p. 31.   6. p. 15.

7. Sowohl übersichtige wie kurzsichtige Augen gestalten sich während der Entwickelungsperiode zum normalgebauten, und solche, die ursprünglich für parallele Strahlen eingestellt waren, sind nach dieser Periode wohl selten für convergirend-, dagegen häufig für divergirend-einfallende Strahlen adaptirt. p. 17.

8. Die Augen von den Kindern in den ersten Lebensjahren zeigen in den einzelnen Fällen ebenfalls eine Verschiedenheit ihrer dioptrischen Einstellung, überwiegend sind sie jedoch im Zustande voller Accommodationsentspannung für grössere Objectabstände adaptirt.

Gleich wie im Allgemeinen die Einstellung für geringere Entfernungen charakteristisch ist dem Auge Neugeborener eben so charakteristisch ist die Einstellung für grössere Entfernungen für das eigentliche Kindesauge. p. 15.

learned from a letter which he has very kindly written to me. He informs me that he did not use atropine in the eyes of the infants or in his own eyes: that he held the lids apart sometimes with his fingers, and sometimes (in case of great resistance on the part of the child), with lid-holders: that the children were laid on their backs upon a bed: that the examinations were made in the upright image, with weak illumination by a Helmholtz ophthalmoscope, consisting of three plane glasses.

In a recent article [*] by Dr. E. Landolt of Paris, he makes the following statement: " Infants are, in the great majority of cases, hypermetropes, even many of those who afterward become emmetropes and myopes." [†] Dr. Landolt informs me, however, that he does not here refer to *newly-born* infants (as he has made no researches among them), but to infants beyond the first years of life; and that he has himself found such to be hypermetropic in the great majority of instances. Even infants of that age Von Jaeger found predominantly myopic. Among children between two and six years old, from an infant school, he found 62 per cent. myopic, and only 8 per cent. hypermetropic: among boys in the country, from six to eleven years old, 43 per cent. myopic, and 11 per cent. hypermetropic; among girls in the country, from five to eleven years old, 56 per cent. myopic, and 10 per cent. hypermetropic.[‡]

My statistics tend to establish the conclusion that emmetropia, myopia, and hypermetropia are all congenital conditions, with a preponderance of hypermetropia. Upon the causes of these differences of adjustment they throw no light. They afford no evidence of a congenital lengthening of the antero-posterior axis of the globe, or a congenital staphyloma posticum. Of course, they contain nothing so valuable as Von Jaeger's measurements on the cadaver.

---

[*] Sur les Causes de l'Amétropie. Communication faite a l'Association Française dans la séance du 27 Août, 1877.

[†] Les enfants sont, dans la grande majorité des cas, hypermétropes, même beaucoup de ceux qui deviennent plus tard emmétropes et myopes.

[‡] As quoted by Donders.

The use of so strong a solution of atropine in such young children may have increased the percentage of hypermetropia, through its supposed flattening effect upon the lens. In the twenty-seven cases measured under the weak solution, ten were myopic and fifteen hypermetropic—a less percentage of hypermetropia ; but the number is too small for any definite conclusion. An attempt was made to examine the whole twenty-seven cases again a week later under the strong solution. Only six could be obtained, however ; these were measured independently, and the former results confirmed in all save one.

If the myopia which was observed in these babies was due merely to the curvature and position of the crystalline lens, it might disappear the next week, or even give place to hypermetropia. At any rate, it would not be the kind of shortsightedness about which we feel anxious, and the finding of it thus early in life would be of no great practical value.

I am fully conscious of the difficulty of attaining perfect accuracy in such investigations as are here reported, and of the fallacy of conclusions drawn from such a small array of statistics. These are simply offered in the hope that they may prove a slight contribution to our knowledge of this subject.

My thanks are due to Sister M. Irene, Sister Superior of the N. Y. Foundling Asylum ; to Dr. C. C. Lee, Dr. W. R. Gillette, Dr. O. D. Pomeroy, and Dr. W. E. Forest, of this city ; to Dr. C. R. Estabrook, Chief of Staff, and Drs. A. B. Farnham, V. G. Culpepper, H. G. Lyttle, J. H. Bryan, J. Habirshaw and L. C. Swift, of the House-staff of Charity Hospital. Without their kind assistance my examinations could not have been made.

A CASE OF POISONING FROM THE USE OF THE COMPOUND TINCTURE OF CINCHONA, PRODUCING PERMANENT CONTRACTION OF THE VISUAL FIELDS AND TEMPORARY IMPAIRMENT OF SIGHT AND HEARING.

By D. B. St. JOHN ROOSA, M.D.

ON the 3d July, 1878, Dr. L. M. Yale asked me to see a case of loss of sight, of which the following history was obtained : Mr. B., æt. 50, a man of very intemperate habits as regards the use of alcohol. He had been accustomed for years to drink enormously of brandy and whiskey at intervals, but there were periods of varying length, from one to three or four months, of total abstinence from intoxicating drinks.

Mr. B. was told that the use of the tincture of cinchona would relieve him from his periodic craving for alcohol. On June 24th of this year he began its use, with a view of correcting his intemperate habits. On that day, as well as on the 25th, 26th, 27th and 28th he continued to take the compound tincture in ounce and two ounce doses, at short intervals, literally drinking it as a beverage from a quart bottle, in which he had caused an apothecary to place as strong a preparation as possible. On the 28th, although he had taken none of his ordinary alcoholic stimulants, his clerk thought from his conduct that Mr. B. had been drinking heavily. Dr. Yale estimates that in these days the patient took an amount of the tincture which would be equivalent to 125* grains of an alkaloid of cinchona. Mr. B. has no recollection of any occurrence after the 27th. He is confident that he took no alcohol, except that contained in the preparation of cinchona, during these days. This, however, may be doubtful, for the clerk of the hotel to which he went when in what proved to be a semi-conscious state on the 28th, states, that while he lay in bed he was constantly ringing the bell for liquor. It is possible that

---

* This amount was afterward found to be nearer 500 grains.

during this time some doses of alcohol were added to those of cinchona, although Mr. B. does not believe this to be the case. On the morning of July 1st he was seen by Dr. Hills in the absence of Dr. Yale. He found the patient stupid or half conscious, with flushed face and conjunctivæ, and apparently unable to see or hear. Mr. B. remembers Dr. Hills' visit on Sunday, and knows that he was then blind and deaf. Dr. Yale saw the patient on Monday and Tuesday, July 2d and 3d. His hearing power improved so much in that time as to become apparently normal, but his vision remained very much impaired. On the day I saw Mr. B., the 3d, he was groping about his room, apparently in excellent general health. $V.\ R.\ E.$=quantitative perception of light. $L.\ E.$ counts fingers at one foot. The ophthalmoscope showed lessened size of the arterial vessels; no abnormity in the veins, lessened number of vessels on the papillæ, but no marked paleness. No changes observed in the membrana tympani. The patient was advised to take strychnia in increasing doses and nutritious diet. On July 6th he was able to walk about. $V.=\frac{20}{30}$ each eye, but the visual fields were very much contracted, so that vision was telescopic.

On July 16, 1878, both visual fields were found concentrically limited. The measurements, drawn on a blackboard 14″ distant, were as follows: Right field, vertical 9 inches; horizontal 7½ inches; limitation most marked on temporal side. Left field, vertical, 7 inches; horizontal, 8 inches; limitation more regular. B. found this symptom rather novel than troublesome. The optic papillæ looked very pale, and the arteries were narrow. July 23d, $V.=\frac{20}{20}$, each eye. Patient states that he can see perfectly well in a straight line, but that when walking about a room he has some difficulty in seeing small articles of furniture.

Sept. 10th.—The same condition is maintained. The strychnia was taken until $\frac{1}{10}$ grain had been reached at a dose, and was continued for two months. The visual field remains as on July 16th.

April 23, 1879.—Mr. B.'s condition remains substantially the same. He continues to abstain entirely from the use of alcohol, and carries on a large business successfully. His vision is still $\frac{20}{20}$ each eye. The visual field has increased somewhat in the left eye. It now measures 9 inches vertically and 16 inches horizontally. F. of $R.\ E.$ 6″ vertically, 9″ horizontally. Limitation most marked at upper-inner quadrant. The optic disks are pale and the arteries small. There are no other ophthalmoscopic appearances.

*Remarks—*Mr. B. had taken no alcohol for some months prior to his beginning the use of the cinchona, and he took none until he became unconscious on the fourth or fifth day. Although he went about and transacted business on the fourth day, he has no recollection of what he did. When found he had an empty bottle (holding a quart) in his room, labelled and giving positive evidence of having contained cinchona. He certainly did not take many drinks, if any, after he reached the hotel, for the clerk, knowing his former habits, and supposing him to be suffering from an ordinary debauch, refused to answer his demands. It is *not known* that he took anything but the cinchona at any time after he began the treatment of the alcohol habit.

We have here, then, a case of hyperæmia of the vessels of the ear from the use of cinchona and alcohol—a hyperæmia which passed away without going on to an exudative process ; but the same condition in the vessels supplying the retina continued until a true vasculitis, with its consequences, resulted.

The future condition of this patient, even if he never assumes the alcohol habit, cannot be regarded without anxiety. It is to be feared that in time the macula may be insufficiently nourished from further contraction of the vessels. The peripheric parts of the retina have now very little, if any, perceptive power ; the nerve is perhaps undergoing atrophy. It is, I think, undoubted from many experiments, among which are my own,* that cinchona causes at least temporary hyperæmia of the vessels of the base of the brain. I am fully aware, however, that, although certainly there was absolutely no loss of sight until the poisoning by cinchona occurred, there may have been changes in his circulation, induced by alcohol, prior to this attack, and I also do not forget that there was enough alcohol in the preparation which he took, to prevent the case from being a typical one of cinchona poisoning, yet the quantity must have been too small to have added much to the effect of the other drug. He may, however, have

---

* Treatise on Diseases of the Ear, 4th edition, page 516.

drank considerable brandy on the day of which he has no recollection, and some also after reaching the hotel. Certain it is, however, that he reached the unconscious state upon doses of the tincture of cinchona alone. Imperfect as is the case in some respects, it may, I think, be regarded as a contribution to our knowledge of the effects of cinchona upon the nutrition of the eye.

THE RELATIONS OF BLEPHARITIS CILIARIS TO AMETROPIA.*

By D. B. St. JOHN ROOSA, M.D.

IT is a well-recognized fact that certain forms of conjunctival inflammation arise from uncorrected errors of refraction. I do not think it is generaly conceded, however, that blepharitis ciliaris often stands in the same relation to ametropia. The principal text-books do not give any prominence to the subject either in the discussion of blepharitis or ametropia. Most, if not all of them, are silent upon the subject. Donders does not, I think, even allude to blepharitis as one of the results of uncorrected strain of the accommodation. In the chapter on Blepharitis in Saemesch's Hand-Buch, by Professor Michel, the subject is not mentioned. Schweigger in his hand-book is also silent upon the point. The same may be said of the treatises of Wecker, Stellwag, and Soelberg Wells. I mention these facts because in speaking of the causal connection of blepharitis with ametropia to some of my professional friends, I found them under the impression that the subject had already been distinctly enough mentioned in the text-books. However much may have been known and said upon the subject in the practice of eye infirmaries, very little has as yet found its way into the literature of ophthalmology.

I therefore present a few statistics as to the connection between diseases of the hair follicles and tarsal glands and

* Read before the International Congress of Ophthalmology, September 1876.

the various forms of ametropia. They are all the cases observed by me in private practice during the last eighteen months. I have attempted to keep a similar record in the Manhattan Eye and Ear Hospital, but there are some omissions in these statistics—that is, the refraction has not been determined in all the cases; I have therefore, not placed them among my private cases. I will say however that, so far as they go, in the opinion of the House Surgeon, Dr. Cheatham, they confirm the results of my own statistics. My conclusions are as follows:—

I. Ametropia seems to be the condition of most eyes affected with blepharitis ciliaris.

II. When the blepharitis is associated with errors of refraction, the cure of the edges of the lids is very much facilitated by and sometimes depends upon correction of the ametropia.

III. Paralysis of the accommodation by the use of atropia will usually, with no other treatment, very much relieve the blepharitis that is associated with ametropia.

IV. Patients suffering from blepharitis associated with ametropia will often ignore any other affection of the eyes than that of the edge of the lids, and deny that they suffer from asthenopia or conjunctivitis, complaining only of the discomfort and disfigurement produced by the disease—and this when the error of refraction is so marked that we would naturally expect quite serious consequences from its non-correction.

V. The form of blepharitis to which my statistics refer is not a mere irritation of the eyelids, such as often accompanies a catarrhal conjunctivitis, but a true hypersecretion of the hair follicles and tarsal glands, attended by the formation of crusts, ulcerated points, and hyperæmia.

VI. Hypermetropia is the error of refraction most frequently associated with blepharitis ciliaris.

I frankly admit that the number of cases I am now able to present does not absolutely prove that blepharitis ciliaris is very frequently caused by ametropia, although I cannot escape the conviction that this is the case. The number is large enough, however, to show a remarkable coincidence at least, and to stimulate others to inquiry in the same direction.

CASE I. Mr. R., æt. 17. Complains of blepharitis, which he has had three or four years. Sometimes has had slight pain in eyes after reading. Accommodation and muscles normal. Refraction, emmetropic. $V=1$.

CASE II. Mr. D., æt. 26. Has had blepharitis and asthenopia for past three years ; complains chiefly of the blepharitis. Has derived no benefit from treatment, which has been from competent surgeons, who have not attempted to prescribe glasses. Refraction, mixed astigmatism, both eyes. Under atropia—

R. E. with $+\frac{1}{12}$ $^{c}[-\frac{1}{12}$ $^{c}V=\frac{20}{30}$.

L. E. " $+\frac{1}{24}$ $^{c}[-\frac{1}{36}$ $^{c}V=\frac{20}{30}$.

Ordered above glasses ; cleansing of the lids with a solution of bicarbonate of soda in water, and the application of red oxide of mercury ointment. Patient reports six months later : Uses eyes with comfort, and has scarcely any blepharitis. Says that redness of the lids returns whenever he leaves off his glasses for a few days. Four months later the lids are entirely well.

CASE III. Miss A., æt. 18. Has had asthenopia and blepharitis since childhood. Accommodation and muscles normal. Refraction, compound hypermetropic astigmatism, both eyes. Under atropia—

R. E. $V=\frac{20}{40}$ with $+\frac{1}{30}\bigcirc+\frac{1}{36}$ $^{c}V=\frac{20}{30}$.

L. E. $V=\frac{20}{40}$ with $+\frac{1}{36}\bigcirc+\frac{1}{36}$ $^{c}V=\frac{20}{30}$.

This patient was freed from the blepharitis, etc., by the glasses.

CASE IV. Miss U., æt. 15. Complains only of blepharitis. Refraction, hypermetropic $\frac{1}{36}$ with each eye. Result of treatment unknown.

CASE V. Mrs. F., æt. 28. Complains of blurring of distant vision, of fatigue in eyes after use, and of blepharitis. Accommodation and muscles normal. Myopia, $\frac{1}{48}$ right eye, $\frac{1}{42}$ left eye ; ordered $\frac{1}{60}$ for both eyes. Four months later reports herself entirely well.

CASE VI. Mr. D., æt. 23. Has had blepharitis and styes for past two years. Some asthenopia for past six months. A. normal. $V=1$. Insufficiency interni, 6° at 12", and 4° at 15'. Refraction, emmetropic. No record of treatment or course.

CASE VII. Mr. V., æt. 28. Complains of blepharitis. $V=1$. Refraction, H. $\frac{1}{48}$, both eyes. After declining glasses for a year nearly, with constant relapses, is now wearing $\frac{1}{60}$ with evident progress in the cure of the blepharitis.

CASE VIII. Mr. W., æt. 28. Blepharitis and asthenopia last two years. A. normal. Refraction, compound myopic astigmatism, both eyes—

$$\text{R. E. with} - \tfrac{1}{48} \subset - \tfrac{1}{24} {}^c V = \tfrac{20}{20}.$$
$$\text{L. E. with} - \tfrac{1}{48} \subset - \tfrac{1}{42} {}^c V = \tfrac{20}{20}.$$

Ordered these glasses. Patient not heard from since.

CASE IX. Mr. D., æt. 28. Complains of blepharitis ; has had it four or five years. Refraction, H. $\frac{1}{36}$, both eyes. $V=1$.

CASE X. Mr. J., æt. 36. Complains of blepharitis, which he has had for several years. A. and muscles normal. Refraction, simple myopic astigmatism $\frac{1}{48}$, both eyes.

CASE XI. Mr. B., æt. 28. Complains of "gritty" sensations about eyes, and blepharitis. Refraction, H. $\frac{1}{21}$, both eyes. $V=1$. One month after, this patient was greatly relieved of his symptoms.

CASE XII. Mr. A., æt. 23. Complains of indistinct vision and of blepharitis. Refraction, simple hypermetropic astigmatism $\frac{1}{18}$, each eye. $V=\frac{20}{50}$. This patient is relieved by the treatment, but a complete cure has not been effected.

CASE XIII. Miss C., æt. 15. Complains of blepharitis. Refraction, hypermetropic, both eyes. Declines to wear glasses.

CASE XIV. Master U., æt. 10. Complains of blepharitis and asthenopia. Refraction, H. $\frac{1}{8}$, each eye. R. E. $V=\frac{20}{30}$. L. E. $V=\frac{20}{50}$. The glasses cause some improvement, but the patient was seen but twice or three times after they were prescribed.

CASE XV. Miss C., æt. 16. Blepharitis since a small child. Treated without success for a year at an eye institution.

$$\text{Refraction, R. E. H.} \tfrac{1}{10}\ V = \tfrac{20}{50} +.$$
$$\text{L. E. H.} \tfrac{1}{24}\ V = \tfrac{20}{60} +.$$

This patient was very much improved, as to the blepharitis, under the use of atropia, in connection with the same treatment that had been previously employed. She passed from observation before she was *entirely* well. Corneal opacities prevented better result from the correction of the hypermetropia, and the glasses had not been ordered when last seen.

CASE XVI. Mr. T., æt. 35. Complains of blepharitis, which he has had since 1858 ; also of asthenopia. A. normal.

$$\text{Refraction, R. E. M.} \tfrac{1}{30}\ V = 1.$$
$$\text{L. E. M.} \tfrac{1}{24}\ V = 1.$$

Insufficiency of recti interni, 7° at 12″. This patient was seen once after glasses were ordered, and was then improved. He had had the usual local treatment for years.

CASE XVII. Mr. A., æt. 24. Complains of having had blepharitis for the past three years ; asthenopia for same period. A. normal. Refraction, M. $\frac{1}{7 5}$, both eyes. $V=1$. Insufficiency interni recti, $5°$ at $12''$. The patient went to Europe before the benefit from glasses could be tested.

CASE XVIII. Mrs. L., æt. 32. Has had asthenopia, slight blepharitis, and muscæ for some time. Unable to do any fine work for past two months. A. and muscles normal. $V=1$. Refraction, emmetropic. This patient is suffering from mental worry, and the eyes but index the whole nervous system. The refraction was tested under atropia.

CASE XIX. Mrs. B., æt. 44. Complains of blepharitis. Refraction, emmetropic, $V=1$. Presbyopia, $\frac{1}{3 6}$.

CASE XX. Mr. B. complains of blepharitis. A. and muscles normal. Refraction very slightly hypermetropic by ophthalmoscope ; not tested with atropia. $V=1$. No note of a second visit.

CASE XXI. Miss L., æt. 21. Complains only of blepharitis, which she had over a year. Refraction, mixed astigmatism, both eyes—

$$\text{R. E. with} - \tfrac{1}{15}\ {}^{c}\ulcorner + \tfrac{1}{42}\ {}^{c}V = \tfrac{20}{30}\ -.$$
$$\text{L. E. with} - \tfrac{1}{15}\ {}^{c}\ulcorner + \tfrac{1}{42}\ {}^{c}V = \tfrac{20}{30}\ -.$$

This patient's blepharitis was considerably improved by the use of atropia for two or three weeks, while the refraction was being tested. There were also evidences of old iritis in her case. She passed from observation immediately after the refraction was determined.

CASE XXII. Mary P., æt. 5. Her mother states that she has had blepharitis for the past eighteen months, and she now has a marked affection of her lids. The refraction is hypermetropic in both eyes $\frac{1}{24}$. On account of the youth of this patient, none but local treatment was advised until she should begin to study.

CASE XXIII. Mr. R., æt. 21. Has had asthenopia and blepharitis of the left eye for the past eighteen months. Blepharitis in right eye for the past three months. A. and muscles normal. Refraction, H. $\frac{1}{3 6}$, both eyes. $V=1$. Ordered $+ \frac{1}{42}$.

CASE XXIV. Master E., æt. $12\frac{1}{2}$. Complains of having pains in his eyes occasionally, and of blepharitis.

Refraction, M. $\frac{1}{45}$, L. E.

M. $\frac{1}{6}$, R. E.

Choroiditis.

CASE XXV. Master F., æt. 6. Has had blepharitis for several months. Refraction, emmetropic by ophthalmoscope. Atropia not used. Local treatment advised.

CASE XXVI. Mr. M., æt. 21. Blepharitis for the past year. Has had a good deal of treatment, but without benefit. A. normal. Insufficiency of internal recti, 4° at 12″. Refraction, M. $\frac{1}{48}$, both eyes.

CASE XXVII. Miss M., æt. 13. Has had blepharitis since a small child. Has been treated frequently, but never permanently cured. Some asthenopia after prolonged use of eyes. Refraction, H. $\frac{1}{30}$, both eyes. $V=1$.

CASE XXVIII. Miss S., æt. 25. Has had blepharitis and asthenopia for five years. A. and muscles normal. Refraction, compound hypermetropic astigmatism, both eyes—

$$\text{R. E. with} + \tfrac{1}{36} \supset + \tfrac{1}{48} {}^c V = \tfrac{20}{30}.$$
$$\text{L. E. } \text{``} + \tfrac{1}{36} \supset + \tfrac{1}{48} {}^c V = \tfrac{20}{30}.$$

Ordered the above glasses. No local treatment for lids. Patient reports four months later that asthenopia is entirely relieved, and that the blepharitis has disappeared.

CASE XXIX. Mr. C., æt. 20. Asthenopia for two years. Quite severe blepharitis for the same period. Refraction, compound hypermetropic astigmatism, both eyes. Under atropia—

$$\text{R. E.} + \tfrac{1}{48} \supset + \tfrac{1}{4}{}^c \text{ axis } 90° \; V = \tfrac{20}{60} +.$$
$$\text{L. E.} + \tfrac{1}{42} \supset + \tfrac{1}{42} \text{ axis } 90° \; V = \tfrac{20}{30}.$$

CASE XXX. Mr. C., æt. 39. Complains of blepharitis. Has had slight asthenopia in the evening, but he is only annoyed by the redness of his lids. Refraction, slightly hypermetropic in both eyes by ophthalmoscope. $\frac{1}{A} = \frac{1}{16}$. Ordered $+ \frac{1}{16}$ for reading. No other treatment. One month later the lids looked better, but not entirely well.

CASE XXXI. Master U., æt. 12. Has had blepharitis for the past four years. He has been treated by the usual remedies, but never cured. Has asthenopia, and a mild form of palpebral conjunctivitis. Refraction, H. $\frac{1}{48}$ each eye, under atropia.

Summary.—Whole number of cases reported, 31.

Complained of blepharitis alone, 15, or about 50 per cent.
Complained of blepharitis and asthenopia, 16.
Cases having refractive error, 26, or $83\frac{9}{10}$ per cent. nearly.
Cases, emmetropic, 9, or about $16\frac{1}{10}$ per cent.

Hypermetropia . .	13
Myopia	. 5
Hypermetropic astigmatism	. 1
Myopic astigmatism . . .	. 1
Compound hypermetropic astigmatism	. 3
" myopic astigmatism	. 1
Mixed astigmatism . .	2
Emmetropia . .	5
	——
	31

The refraction in these thirty-one cases was tested under atropia whenever it was allowed. I cannot accept statistics on this subject for myself, that are not made up in this way. For I am led to believe, from some considerable observation, that even experienced and competent observers sometimes declare an eye emmetropic which they have examined with the ophthalmoscope, without atropia, when the use of the mydriatic will show hypermetropia of more than a sixtieth. If every one of my cases had been tested under atropia, the percentage of hypermetropia would perhaps have been increased.

THE RELATIONS OF BLEPHARITIS CILIARIS TO AMETROPIA.

By D. B. ST. JOHN ROOSA, M. D.*

At the Fifth International Ophthalmological Congress, held in New York in 1876, I read a paper on the above subject, and stated the following conclusions, as those which seemed to me to be deduced from my cases:

1. Ametropia seems to be the condition of most eyes effected with blepharitis ciliaris.

2. When the blepharitis is associated with errors of refraction, the cure of the edge of the lids is very much facilitated by, and sometimes depends upon, correction of the ametropia.

3. Paralysis of the accommodation by the use of atropia will usually, with no other treatment, very much relieve the blepharitis that is associated with ametropia.

4. Patients suffering from blepharitis that is associated with ametropia will often ignore any other affection of the eyes than that of the edge of the lids, and deny that they suffer from asthenopia or conjunctivitis, complaining only of the discomfort and disfigurement produced by the disease; and this when the error of refraction is so marked that we would naturally expect quite serious consequences from its non-correction.

5. The form of blepharitis to which my statistics refer is not a mere irritation of the edge of the lids, such as often accompanies a catarrhal conjunctivitis, but a true hypersecretion of the tarsal glands and hair-follicles, with the

---

* Read before the American Ophthalmological Society, 1878.

formation of crusts, and sometimes the development of ulceration.

6. Hypermetropia is the error of refraction most frequently associated with blepharitis ciliaris.

I have here restated them, because one writer* has not, I think, kept them sufficiently in view in his paper intended as an answer to mine, on the relation of ametropia to blepharitis ciliaris. Since reading my own paper I have continued my investigations in this and a cognate subject, and I now beg the indulgence of the Society for a few additional statistics, as well as for a brief reply to some of the objections that have been made to my conclusions.

Two hundred and one cases of blepharitis cilliaris have been observed at the Manhattan Eye and Ear Hospital since the reading of my paper.

In spite of my efforts to secure an examination of the refractive state of the eyes thus affected, in only forty-eight of these was it noted. In these the refractive state is recorded as follows:

> Hypermetropia 34
> Myopia 1
> Astigmatism . . . . '. . . . 7
> Emmetropia 6

Hypermetropia.

Case   1. (Under atropia), $+ \frac{1}{30}$.
"   2. By ophthalmoscope.
"   3. $+ \frac{1}{30}$.
"   4. Hypermetropia cum Presbyopia.
"   5. By ophthalmoscope, $\frac{1}{16}$.
"   6. "   "
"   7. "   "   R.E., $\frac{1}{40}$; L.E., $\frac{1}{24}$.
"   8. (Under atropia), R.E., $\frac{1}{20}$; L., $\frac{1}{24}$.
"   9. "   R.E., $\frac{1}{30}$; L.E., E.
"   10. Hypermetropia.
"   11. "
"   12. " $\frac{1}{40}$.
"   13. "

---

* F. C. Hotz, *Chicago Medical Journal and Examiner*, April, 1878.

" 14. Atropia, R,, $\frac{1}{24}$; L., $\frac{1}{30}$.
" 15. H., $\frac{1}{30}$.
" 16. Ophthalmoscope.
" 17. R., $\frac{1}{60}$; L., $\frac{1}{30}$.
" 18. Atropia, R.E., $\frac{1}{30}$; L., $\frac{1}{30}$.
" 19. Ophthalmoscope.
" 20.   " $\frac{1}{20}$.
" 21.   "
" 22.   "
" 23. R.E. (E.) ; L., $\frac{1}{30}$.
" 24. $\frac{1}{24}$.
" 25. R., $\frac{1}{50}$; L., $\frac{1}{60}$.
" 26. $\frac{1}{40}$.
" 27. Ophthalmoscope, R.E., $\frac{1}{14}$; L.E., $\frac{1}{16}$.
" 28. H. and strabismus.
" 29. Ophthalmoscope.
" 30. $\frac{1}{30}$; ophthalmoscope, $\frac{1}{16}$.
" 31. H.,   "
" 32. H.,   "
" 33. H.
" 34. $\frac{1}{20}$.

Myopia.

Case 1. V.R.E., $\frac{20}{60}$; L.E., $\frac{20}{30}$, with — $\frac{1}{12}$.

Astigmatism.

Case 1. R.E., — $\frac{1}{36}$ ⊃ — $\frac{1}{48}$ axis 180°; L.E., — $\frac{1}{36}$ — $\frac{1}{48}$, 180°.
" 2. R.E., + $\frac{1}{20}$c axis 90°; L.E., + $\frac{1}{18}$; A., 90°.
" 3. Astigmatism.
" 4. Mixed astigmatism.
" 5. R.E., $\frac{20}{40}$; L.E., $\frac{20}{40}$ — $\frac{1}{12}$ c., 180°.
" 6. R.E., + $\frac{1}{12}$c axis 90 ⊃ — $\frac{1}{30}$c; L.E., $\frac{20}{30}$ — $\frac{1}{18}$c., 15°.
" 7. R.E., + $\frac{1}{80}$s ⊃ + $\frac{1}{14}$c 90°; L.E., V, $\frac{20}{30}$ E. (?).

Emmetropia.

Case 1. V. = $\frac{20}{40}$. (No reason given in register for amblyopia.)
" 2. E.
" 3. R.E., $\frac{20}{40}$; L.E., $\frac{20}{40}$ +.
" 4. V. = $\frac{20}{30}$.
" 5. V. = $\frac{20}{20}$.
" 6. E.R.E. Left opacity of cornea.

Opacities of the Cornea.—There were eight cases of opacities of the cornea among those of which the refraction was not tested.

There was also one case of diminution of the acuteness of vision, with no assigned cause for the loss. R.E., $\frac{20}{40}$ +; L. E., $\frac{20}{70}$. Not improved by glasses.

This increases the number of cases that may be said to have been examined to 57.

Even a superficial examination of these statistics shows that some of the statements made in my first paper are sustained by them. There is even a larger proportion of refractive error among them than in the first series. This ametropia is also generally of such a degree as to require correction. That such a correction will in many cases assist in the cure of the blepharitis will hardly be denied. It is a step far beyond this, I admit to say that the blepharitis was caused by the strain from an uncorrected error of refraction. I am not sure, judging from my own experience in another direction, but that we shall be obliged to modify our views as to how injurious is the strain of accommodation in hypermetropia; but I think that we may still believe that in many cases uncorrected hypermetropia will produce all the consequences of continued hyperæmia of the edges of the lids.

Since the publication of my paper I have seen 40 cases of blepharitis ciliaris in private practice, as follows;

CASE I.—Miss P., æt. 37. Is subject to styes; always has had asthenopia; now has blepharitis. R.E., $\frac{20}{30}$; H., $\frac{1}{18}$; V. = $\frac{20}{30}$. L.E., $\frac{20}{200}$. No improvement from glasses. Under atropia, H. R.E., = $\frac{1}{18}$; L.E., + $\frac{1}{20}$ $\subset$ $\frac{1}{24}$c 90°; V. = $\frac{20}{40}$. Glasses were prescribed, and two months after the patient writes that the glasses have given her "full satisfaction." There is no account as to the styes or blepharitis.

CASE II.—Æt. 17. About eleven months ago the patient observed that his distant vision was dim; has had asthenopia and blepharitis since then. M. = $\frac{1}{18}$: R.E., L.E., $\frac{1}{30}$.

CASE III.—C. C., æt. 13. Has had blepharitis and asthenopia for about a year. V. = $\frac{20}{80}$. Under atropia, R.E., + $\frac{1}{60}$c 90° V. = $\frac{20}{16}$. L.E., + $\frac{1}{60}$c 90° V. = $\frac{20}{16}$. V. $\frac{20}{40}$ under atropia without glasses. Four months after the patient was using his eyes with comfort.

CASE IV.—F. G., æt. 28. Has asthenopia in the evening and blepharitis constantly. Under ophthalmoscope, eyes are H., rejects glasses. V. = $\frac{20}{20}$. Local treatment alone employed, and one month after the patient was better.

CASE V.—C. H. G., æt. 19. 'Has had blepharitis for eight or nine years, also asthenopia. V. = $\frac{20}{20}$. Under atropia, H. = $\frac{1}{12}$.

CASE VI.—Mrs. B., æt. 42. Has always suffered from asthenopia. Has blepharitis, and there is a chalzion on the right upper lid. H. $\frac{1}{48}$ and $\frac{1}{12}$. There is also marked insufficiency of the interni.

CASE VII.—P. T., æt. 20. The eyelids have been inflamed for two years ; asthenopia in the evening ; H. by opthalmoscope. Rejects all glasses for the right eye, accepts $+ \frac{1}{72}$ for the left, under atropia $+ \frac{1}{12}$. In a few days there was manifest H. of $\frac{1}{48}$.

CASE VIII.—A. C. B., æt. 25. Asthenopia, blepharitis in left eye for two weeks ; under atropia H = $\frac{1}{36}$.

CASE IX.—Mr. S., æt. 40. Has had red eyelids (edges) as long as he can remember. Sometimes, not often, has asthenopia. V. = $\frac{20}{20}$. Under atropia, R.E.H. = $\frac{1}{48}$; L.E.H. = $\frac{1}{12}$.

CASE X.—D. G. B., æt. 38. Has suffered from redness of the edges of the lids for the past five or six weeks. 'Has never had asthenopia. V. = $\frac{20}{20} - \frac{20}{20}$, with $- \frac{1}{60}$. With ophthalmoscope eyes seem to be H. Ordered $+ \frac{1}{48}$.

CASE XI.—Mr. C., æt. 31. Blepharitis ever since he can remember. No asthenopia. R.E., $\frac{20}{30} - $ L.E., $\frac{20}{20}$. Rejects glasses. Ophthalmoscope shows H. $\frac{1}{24}$ R.E.; L.E., $\frac{1}{48}$. Under atropia, R.E., $+ \frac{1}{48} \subset \frac{1}{48}^c$ 90° $\frac{20}{20}$; L.E., $+ \frac{1}{48} \subset \frac{1}{60}^c$ 90° $\frac{20}{20}$.

CASE XII.—Miss C. A. S., æt. 17. Asthenopia in the evening for a year. Red line along the edge of the lids for one year. V. = $\frac{20}{20}$. Under atropia, H. = $\frac{1}{36}$.

CASE XIII.—Dr. B. Has had blepharitis since he was seven years old. V. = $\frac{20}{20}$. Rejects glasses. H. as with ophthalmoscope ; atropia, R.E., $+ \frac{1}{60}^c$ 80°. V. = $\frac{20}{20}$; L.E., $+ \frac{1}{60}^c$ 90° $\frac{20}{20}$.

CASE XIV.—I. B., æt. 38. Asthenopia for the past four or five years. Has worn glasses, but latterly they have not afforded relief. Redness of the edge of the lids. R.E.H. $\frac{1}{48}$; L., $\frac{1}{36}$ — manifest.

CASE XV.—Miss H., æt. 18. Always has had asthenopia, and the eyelids get red. V. = $\frac{20}{20}$. Under atropia, R.E., $\frac{1}{12}$; L. E., $\frac{1}{48}$.

CASE XVI.—M. M., æt. 7. Eyes "weak" for a year. Slight blepharitis. V. = $\frac{20}{20}$. Advised atropia. No record of ophthalmoscope examination.

CASE XVII.—F. G. D., æt. 19. Asthenopia and slight blepharitis. V. = $\frac{20}{20}$. Under atropia, H. = $\frac{1}{48}$. Two months after, still has asthenopia and inflamed lids.

CASE XVIII.—Mrs. J., æt. 36. Has had asthenopia since she was eight or nine years old ; blepharitis for a year. Under atropia, + $\frac{1}{30}$. Ordered glasses, and two months after writes that she uses them with perfect comfort.

CASE XIX.—Miss B., aet. 22. Redness of the edges of the lids as long as she can remember. V. = $\frac{20}{20}$ rejects glasses. Under atropia, R.E., H. $\frac{1}{30}$; L., $\frac{1}{36}$. Two months after, the patient was doing well.

CASE XX.—E. A., æt. 24. Has had sore eyes ever since birth. Has entropion and blepharitis. V. = $\frac{20}{40}$ + R.E.; L.E., $\frac{20}{20}$. Under atropia, R.E., + $\frac{1}{30}$ c 90° $\frac{20}{30}$, L.E., + $\frac{1}{20}$ c 120° $\frac{20}{20}$. A year after has worn glasses, and eyelids are much better. An operation for entropion was performed, which relieved it, but the blepharitis remained, when the glasses were ordered.

CASE XXI.—Mrs. A., æt. 23. Eyelids have always been "diseased." Subject to styes. R.E., $\frac{20}{30}$ — L.E., $\frac{20}{20}$. Manifest H. = $\frac{1}{42}$.

CASE XXII.—Miss L., æt. 20. Blepharitis, phlyctenular keratitis. By ophthalmoscope, H. astigmatism. No improvement by glasses.

CASE XXIII.—L. W. D., æt. 14. Lachrymal catarrh, blepharitis. H. by ophthalmoscope.

CASE XXIV.—R. E., æt, 21. Inflammation of the edge of the lids ever since childhood ; no asthenopia. R.E., central corneal opacities. L.E., H. = $\frac{1}{20}$.

CASE XXV.—S. J. F., æt. 21. Has been troubled with redness of the edges of the lids for four or five years. Asthenopia under atropia, R.E., + $\frac{1}{48}$ c 60° — $\frac{20}{20}$; L.E., + $\frac{1}{48}$ 120° $\frac{20}{20}$.

CASE XXVI.—J. V. W., æt. 22. Eyes "weak" for several years. Lids have been red for several months. R.E., $\frac{20}{25}$; under atropia, L.E., $\frac{20}{20}$. R.E., H. $\frac{1}{36}$; L.E., + $\frac{1}{30}$ C $\frac{1}{48}$ c 90°.

CASE XXVII.—F. G., æt. 22 (?). Blepharitis. Is wearing — $\frac{1}{18}$, selected by himself. V. = $\frac{20}{20}$ with — $\frac{1}{30}$. These were substituted for — $\frac{1}{18}$; three months after lids were much better.

CASE XXVIII.—S. O., æt. 20. Styes for four or five years.

Has trachoma and blepharitis. V. = $\frac{2}{3}\frac{0}{0}$. Rejects glasses. Under atropia, H. = $\frac{1}{30}$; V. = $\frac{2}{2}\frac{0}{0}$.

CASE XXIX.—H. D. N., æt. 21. Weak eyes since he was three years old. Blepharitis; accepts + $\frac{1}{30}$. Under atropia, R. E., $\frac{1}{18}$; L.E., + $\frac{1}{24}$. The treatment was of benefit in this case, as stated by patient five months after.

CASE XXX.—B. A. C., æt. 22. Asthenopia, trachoma, and blepharitis. H.M., $\frac{1}{42}$. Declines glasses. Five months after, blepharitis no better.

CASE XXXI.—F. B., æt. 20. For two years has had blepharitis, and has been unable to read at night. R.E., E.; L.E., mixed astigmatism.

CASE XXXII.—E. B., æt. 14. Has had blepharitis for eighteen months, asthenopia for a year. H. m. = $\frac{1}{30}$.

CASE XXXIII.—Jennie Y., æt. 11. Has always had blepharitis. V. = $\frac{2}{2}\frac{0}{8}$ — H. m. $\frac{1}{48}$, and V. = $\frac{2}{2}\frac{0}{0}$.

CASE XXXIV.—Mrs. F. B., æt. 27. Blepharitis for six months. R.E., $\frac{2}{6}\frac{0}{0}$; L.E., $\frac{2}{2}\frac{0}{0}$. Under atropia, R.E., + $\frac{1}{48}$ ⊃ $\frac{1}{30}$ᶜ $\frac{2}{2}\frac{0}{0}$; L.E., $\frac{1}{30}$.

CASE XXXV.—Mrs. A. B. De, æt. 29. Blepharitis and asthenopia for four or five years. Has used local applications without benefit, under advice of an oculist. V., $\frac{2}{2}\frac{0}{0}$. Rejects all glasses. Under atropia, + $\frac{1}{36}$ᶜ; axis, 90°, each eye.

After use of glasses patient is now, six months after, quite well as to lids; has occasional attacks of asthenopia. The use of atropia did great good to the lids without glasses.

CASE XXXVI.—Mrs. A. E., æt. 41. Has had blepharitis and asthenopia for twenty-five years.

Oph. shows H.; advised + $\frac{1}{36}$, and patient, who is often seen, states that her eyes are well.

CASE XXXVII.—Miss M. O. R., æt. 21. Has always had blepharitis. Under atropia, H.R.; E., + $\frac{1}{48}$; L.E., + $\frac{1}{48}$ᶜ 180°.

CASE XXXVIII.—C. L. W., æt. 11. Blepharitis for one year. Atropia, R.E., $\frac{1}{48}$; L., $\frac{1}{42}$. The patient did well under use of glasses.

CASE XXXIX.—R. D. M., æt. 19. Pain in eyes; blepharitis for several years. Under atropia, R.E., — 60ᶜ 180° $\frac{2}{2}\frac{0}{0}$; $\frac{2}{6}\frac{0}{6}$ without glasses. L.E., + $\frac{1}{60}$ᶜ 180°; $\frac{2}{3}\frac{0}{0}$ without glasses.

CASE XL.—Henrietta L. Has blepharitis; has been wearing glasses for some time in the evening. Under atropia, R.E., + $\frac{1}{4}$ ⊃ + $\frac{1}{48}$ᶜ 90° $\frac{2}{2}\frac{0}{0}$; L.E., + $\frac{1}{42}$ ⊃ + $\frac{1}{35}$ 90°.

CASE XLI.—J. M. C., æt. 22. Blepharitis for ten years; pho-

tophobia; no asthenopia; trachoma. V. $= \frac{20}{40} - 35$. R.E., —
$\frac{1}{60} \supset - \frac{1}{60}^c \ 90° \frac{20}{30}$; L.E., $- \frac{1}{60} \supset - \frac{1}{60}^c \ 90° \frac{20}{30}$.
CASE XLII.—Miss P., æt. 14. Blepharitis and asthenopia.
V. $\frac{20}{30}$ — R.E.; L.E., $\frac{20}{40}$; atropia, R.E., $+ \frac{1}{30}$; V. $= \frac{20}{20}$. Patient
did well.
CASE XLIII.—Mrs. E., æt. 32. Asthenopia always; blephar-
itis. Atropia, R.E., $- \frac{1}{60}^c$, axis 90°; L.E., E.
CASE XLIV.—Bella P. Blepharitis; hyperopia, $\frac{1}{60}$.
CASE XLV.—M. N., æt. 7. Asthenopia and blepharitis. H.
by ophthalmoscope, $\frac{1}{48}$.
CASE XLVI.—Mrs. C. B., æt. 43. Asthenopia; blepharitis.
H. $= \frac{1}{36}$.
CASE XLVII.—J. W., æt. 9. Asthenopia; slight blepharitis;
refraction could not be determined.

Summary.—Total number of cases seen in private practice:
Hypermetropia.—Refraction the same in eye. Degree — $\frac{1}{36}$, $\frac{1}{48}$,
$\frac{1}{60}$, $\frac{1}{30}$, $\frac{1}{48}$, $\frac{1}{36}$, $\frac{1}{48}$, $\frac{1}{42}$, $\frac{1}{36}$, $\frac{1}{42}$, $\frac{1}{30}$, $\frac{1}{48}$, $\frac{1}{42}$, $\frac{1}{36}$, $\frac{1}{48}$, $\frac{1}{36}$, $\frac{1}{42}$, $\frac{1}{42}$.
When the eyes were of different refraction—R.E., $\frac{1}{48}$, L.E., $\frac{1}{42}$;
R.E., $\frac{1}{30}$, L.E., $\frac{1}{36}$; R.E., $\frac{1}{48}$, L.E., $\frac{1}{30}$; R.E., $\frac{1}{24}$, L.E., $\frac{1}{48}$; R.E.,
$\frac{1}{48}$, L.E., $\frac{1}{42}$; R.E., $\frac{1}{48}$, L.E., $\frac{1}{42}$.
Degree undetermined—1.
Total number of cases of H.—25.
Hypermetropic astigmatism.—The same in both eyes, $\frac{1}{60}$, 180°,
$\frac{1}{36}$, axis 90°; R.E., $\frac{1}{48}$, a. 60°; L. E., $\frac{1}{48}$, a. 120°; R.E., $\frac{1}{60}$ a. 90°;
L.E., $\frac{1}{60}$, a. 90° $\frac{1}{60}$, a. 90°—5.
Of different degrees—R.E., $\frac{1}{60}$; L.E., E.; R.E., $\frac{1}{4} \supset \frac{1}{42}^c$ 90°;
L.E., $\frac{1}{42} \supset \frac{1}{36}$ 90°; R.E., $\frac{1}{48} \supset \frac{1}{36}^c$; L.E., $\frac{1}{36}$ H.; R.E., $\frac{1}{36}$; L.E.,
$\frac{1}{36} \supset \frac{1}{48}^c$; R.E., $\frac{1}{48}$, axis 60°; L.E., $\frac{1}{48}^c$, axis 120°; R.E., $\frac{1}{30}$ 90°;
L.E., $\frac{1}{20}^c$ 120°; R.E., $\frac{1}{60}$ 80°; L.E., $\frac{1}{60}$ 90°; R.E., $\frac{1}{48} \supset \frac{1}{48}^c$ 90°;
L.E., $\frac{1}{48} \supset \frac{1}{60}^c$; R.E., $\frac{1}{18}$; L., $\frac{1}{20} \supset \frac{1}{24}^c$ 90°—9.
Degree undetermined, 1. Total, 15.
Myopic astigmatism.—$\frac{1}{60} - \frac{1}{60}^c$ 90°—1.
Myopia.—$\frac{1}{30}$; R.E., $\frac{1}{18}$; L., $\frac{1}{30}$—2.
Mixed astigmatism.—R.E., E.; L., mixed astigmatism—1.
Refraction not determined—2.

These statistics are certainly very different from those
given by Dr. A. Alt,[*] who examined forty-eight cases of
blepharitis with a view to test the connection between ame-

---

[*] *Archives of Ophthalmology and Otology,* vol. vi., p. 180.

tropia and inflammation of the lids. According to to Dr.
Alt, "thirty-nine of them had emmetropia, five myopia,
three hyperopia, one astigmatism." I know of no way of
reconciling Dr. Alt's statistics with my own, since there is
no record of the manner in which the refraction was esti-
mated, or of what Dr. A. considers an emmetropic eye. I
do not regard any eye as emmetropic, which not having
V. $\frac{20}{20}$, obtains it under the influence of atropia with a convex
glass of $\frac{1}{80}$ or upward ; neither do I regard the test by the
ophthalmoscope as sufficient to determine the existence of
latent hypermetropia.

Dr. F. C. Hotz* examined eighteen cases in private
practice, " of which five, or thirty-three per cent., showed
ametropia (four showed hypermetropia, one myopia, and
one astigmatism [6 ?]). Dr. H. did not use atropia for the
determination of the refraction in his cases ; and he con-
cedes that, had he employed it, some of his cases might
have shown a slight degree of hypermetropia. Dr.
Hotz is of the opinion that, however great may be the
proportion of ametropia, there is no etiological connection
with this and blepharitis. Dr. H. is mistaken, I think, in
his statement that Dr. Erisman examined his cases under
atropia. The mistake is all the worse because a correct
account of Erisman's statistics rather makes for Hotz's
v ews than against them. Erisman† states that under
atropia it is probable that there would be a larger percentage
of H. than he found and that the hyperopic eye is probably
the normal one in youth.

Dr. Hotz assumes that no strain on the accommodation
occurs in myopic eyes unless they are armed with unsuitable
glasses. I believe that not only those that have improper
glasses, but also those myopes that wear no glasses at all,
suffer from strain.

Dr. Hotz's argument against the occurrence of blepharitis
without asthenopia may be answered by my own experience
and that of my associate in practice, Dr. E. T. Ely, as
follows : We have seen many cases of blepharitis without

---

* *Chicago Medical Journal,* April, 1878.
† *Graefe's Archiv,* B. 17, A. I.

noticeable asthenopia. Some of these were cured for the time by a simple paralysis of the accommodation by means of atropia, and many are the cases that have been treated for years without apparent benefit, which were entirely relieved of their unsightliness and discomfort after glasses were worn in conjunction with a simple local treatment.

There is no probable way of accounting for the cure in these cases, except by a reference to a correction of the error of refraction. In saying this, I am very far from asserting that every case of blepharitis is caused by an error of refraction.

I cannot agree with those who have objected to my views, who argue that slight degrees of H. seldom give rise to asthenopia. My experience is just the other way. Of course I consider a $\frac{1}{42}$ a low degree; perhaps a $\frac{1}{24}$ is the first point at which H. may be said to be at all of a high degree; certainly there is often great relief from the correction of a $\frac{1}{48}$ or even a $\frac{1}{60}$, when that sixtieth is in one meridian only, while the other is emmetropic.

Donders regards the degrees of H. from $\frac{1}{100}$ to $\frac{1}{40}$ as not to be observed in youth. Under ordinary circumstances I grant this, but in some cases I am inclined to the view that $\frac{1}{48}$ may need correction after ten years of age, at least for a time.

In others, as I have shown in another paper, a $\frac{1}{24}$ may exist even in the eyes of a student, and never cause any inconvenience of any kind. When all the factors are at work that produce asthenopia, blepharitis, and so forth, even a very slight error of refraction will materially assist in making the patient uncomfortable, and its correction will do great good.

Dr. Hotz is of the opinion that blepharitis occurs chiefly in children. I am not sure that this is a fact; certainly those cases observed by me in private practice were chiefly among adults. Of two hundred and one cases seen at the Manhattan Eye and Ear Hospital, a little more than fifty per cent. occured in persons under fifteen; but of very young children, that is, of five years of age or under, there was only twenty-eight per cent., where-as Dr. H. states that

he has observed the greatest number among these. Dr. H.
asks if I omitted young children from statistics, because I
could not employ the tests of vision. I answer that I have
inserted in my statistics every case that has presented itself
to me. Because children do not read and write, it cannot
be argued, however, that they do not use the ciliary muscle
and interni. Any one who has watched a child at play, for
instance, picking up small objects for a number of minutes
at a time, will soon be convinced that they are often using
a great deal of accommodative power, in their effort for
exact vision, before they learn to read and write.

If Dr. Alt and Dr. Hotz will examine their cases under
atropia, I am sure they will form different conclusions, as to
the existence of ametropia in connection with blepharitis
ciliaris, from those that they have expressed. I still believe
that there is much more than mere coincidence, in the frequent occurrence of blepharitis in connection with strain
on the accommodation from refractive defects and opacities
of the cornea, and that correction of the error will do as
much toward the relief of the hyperæmia and inflammation of the lids as the correction of hypermetropia does
for asthenopia. I suspect, however, that since Donders'
exposition of asthenopia we have all overrated the curative
power of glasses, and that some writers have been led
greatly into error in ascribing nervous affections too exclu-
sively to the influences of an uncorrected error of refraction.
Certain it is, as I have before indicated, that there are many
cases where quite a high degree of uncorrected hypermetropia does no harm to the subject of it.

AN EXAMINATION UNDER ATROPINE, OF THE RE-FRACTIVE STATE OF EYES WITH NORMAL VISION (²⁰⁄₂₀), AND WHICH HAD NEVER BEEN AFFECTED WITH ASTHENOPIA OR INFLAMMATION,

By D. B. ST. JOHN ROOSA, M. D.*

IT will be at once seen that it is a very difficult thing to procure for examination many eyes that answer to the above-named conditions; and when is added to them another, that is, that the subjects shall be between twenty and forty years of age, the difficulties are not diminished. The importance of such an examination is, however, I think, considerable, for it has been claimed by some that latent hypermetropia and astigmatism are the causes not only of asthenopia, conjunctivitis, blepharitis, etc., but also of chorea and lachrymal disease, while on the other hand, as good an authority as Dr. E. Hansen, of Copenhagen, is very scepti-cal as to the universal or even general value of the correc-tion of hypermetropia for the relief of asthenopia. In a conversation with Dr. Hansen, last summer, he informed me that he found that glasses in very many instances seemed to utterly fail to afford relief to asthenopia associ-ated with and apparently caused by hypermetropia. Dr. Hansen has peculiar opportunities for observation upon this subject, for he enjoys almost a monopoly in ophthalmic practice in the Danish capital; and patients have not the same liberty that is enjoyed in American, English or German cities, of roaming about from one oculist to another when their glasses do not suit, or their eyes do not get well. We have had abundant statistics as to the existence of myopia

* Read before the American Ophthalmological Society, 1878.

and as to the apparent refraction of unatropinized eyes. I have thought it desirable, however, to learn the refraction of those that have been classified as emmetropic. It is claimed, however, that there is a source of error in drawing conclusions from this kind of an examination. It has been said that we almost invariably flatten the lens in putting an eye under the full influence of atropia. If this be true, we always either diminish myopia under such circumstances, or convert emmetropia into hypermetropia.

My statistics show that there was a certain proportion of cases among those examined that continued to reject even the weakest convex glasses, after atropia had been thoroughly used. Besides, I think every one of us had seen cases of myopia which were not at all reduced in degree, after the drug had been employed long enough to test the matter. I cannot, in the face of these facts, assent to the view that atropia flattens the lens except in so far as it acts by paralyzing the ciliary muscle, and thus brings out the true length of the eyeball in a state of rest. Apart from this view, however, there is another importance to such an examination as I have made.

It has been generally, if not universally conceded, that, if atropia reveals hypermetropia in case of inability to use the eyes for continued work, glasses are to be given somewhat in accordance with those taken when the eyes are atropinized. Some authorities even advise full correction. Now, if we find that a large proportion of eyes that have never suffered from asthenopia take convex glasses which were rejected before the mydriatic was used, we may suspect that the mere existence of lessened refraction or latent hypermetropia is not positive proof that asthenopic eyes are to be cured by positive glasses—in other words, that there be other factors producing the trouble. The amount of latent hypermetropia revealed in these cases thus rigorously selected is in many instances equal to that for which oculists all over the world, ever since the publication of Donders' book have been prescribing glasses. Indeed, I think it has been pretty generally assumed that the cause for a given case of asthenopia had been found, if atropine revealed hypermetropia. In later times it has been claimed that the etiology of headaches, of chorea, of lachrymal disease, has been very largely traced to latent and uncorrected refractive errors. Whatever the subjoined statistics may be worth, they at least show, few as they are, that a decided amount of latent hypermetropia is entirely consistent with uninterrupted and painless use of the eyes. This, however, is a fact that I commented upon some two years since, having deduced it from other grounds, namely, that we were quite often called upon to fit presbyopes who were also hypermetropes, with glasses for the first time when their accommodation began to fail, they never before in all their lives having had any necessity for glasses.

We cannot any longer assume, I think, that latent hypermetropia is necessarily the sole factor in the production of asthenopia, much less of troubles of the head and nervous system. That hypermetropia does often produce many of these things, and that its correction will often relieve them. I am of course very far from denying. But I am unable to say why it does not in all of them, especially when in looking over my statistics I find that, in some of the cases that have never suffered for an instant from asthenopia, all the conditions for the injurious influences of strain of the accommodation have existed in feeble organizations and weak muscular systems.

Had I not had the facilities afforded by a large class of medical students, who are very ready for physiological experiments, I should never have succeeded in getting the cases that I now present. I must here present my grateful acknowledgements to the members of the Class of 1877–78 in the University of the City of New York, who so kindly submitted to the troublesome tests. It of course goes without saying, that all the cases now about to be quoted conform to the conditions stated in the foregoing, *i. e.* the patients were not presbyopic ; they had also passed the age during which, in the opinion of some authorities, hypermetropia is always found ; they had V. $\frac{20}{20}$, the sight was blured by convex glasses of $+ \frac{1}{60}$, and in some cases of still weaker power, they had never in all their lives suffered from asthen opia or ocular inflammation, and their accommodation was paralyzed by a four-grain solution of sulphate of atropia.

Observations.

I.—Dr. S., æt. 30. V. R. E. = $\frac{20}{13}$; L. E., $\frac{20}{20}$. After a four-grain solution of the sulphate of atropia had been used four times in twenty-four hours, the vision of the right eye was reduced to $\frac{20}{50}$. It became $\frac{20}{13}$ with $+\frac{1}{60}\bigcirc\frac{1c}{60}$ 90°. This gentleman used atropia for four days, and had some marked constitutional symptoms. The final examination showed H. $\frac{1}{30}$ with H. as $+\frac{1}{60}$ axis 90°.

II.—J. C. M., æt. 21. Began to go to school at seven, and has used his eyes as a student ever since ; V. = $\frac{20}{13}$ +. All positive glasses blur. After the use of atropia, gr. four solution, three times in two days, vision with the right eye became $\frac{20}{20}$ with $+\frac{1}{42}$; L. E., $\frac{20}{40}$ without glasses, and $\frac{20}{20}$ with $+\frac{1}{30}$.

III.—W. H., æt. 34. After atropia four times in twenty-four hours, L. E. having been atropinized, V.=$\frac{20}{20}$ + before atropia, $\frac{20}{30}$ with $+\frac{1}{60}$; two days after, $\frac{20}{20}$ — with $+\frac{1c}{60}$ 90° $\frac{20}{20}$.

IV.—H. J. H., æt. 29. Has used his eyes as a student and teacher since he was six or seven years old ; is thin and pale, only moderately well developed ; right eye examined after atropia instillation twice a day for two days. Accepts $+\frac{1}{24}$ on the second day, the day after $+\frac{1}{30}$ with which V.=$\frac{20}{20}$. Left eye of same patient was tested in the same way a month before ; vision was $\frac{20}{70}$ under atropia without a glass, and $\frac{20}{20}$ with $+\frac{1}{24}$. The subject is anæmic and thin ; has bronchitis every winter ; one member of family died of phthisis.

V.—W. H., æt. 32. Has not been a very constant student ; never the slightest trouble with his eyes, except occasionally after reading very fine print by gaslight for several hours, eyes have felt tired and he has been inclined to rub them ; never obliged to stop work on account of his eyes. Is in good health and of robust appearance. After the use of atropia three times a day for three days, V. = $\frac{20}{20}$ and all glasses are rejected.

VI.—W. F. C., æt. 27. School at an early age ; has studied medicine for a year. Interval of ten years, when his eyes were used only in ordinary reading. "Fatigue" of eyes after three hours' work. Is robust and in good health, V. = $\frac{20}{20}$ R. E., accepts $+\frac{1}{72}$, rejects $+\frac{1}{60}$; L. E., $\frac{20}{20}+\frac{1}{72}$, blurs ; atropia two days, L. E. V.=$\frac{20}{70}$; with $+\frac{1}{30}$. = $\frac{20}{20}$.

VII.—H. R. D., æt. 21. At school until eighteen ; writing for a year after; has studied medicine since. V. = $\frac{20}{20}$ + $\frac{1}{12}$, blurs; after one use of atropia, V. was not $\frac{20}{20}$ unless with + $\frac{1}{30}$.

VIII.—R. D. B., æt. 22. Has always been a student ;=$\frac{20}{20}$, rejects + $\frac{1}{80}$. Atropia for three days, V. = $\frac{20}{20}$, declines + $\frac{1}{80}$.

IX.—C. A. V. R., æt. 22. At school regularly in Germany from sixth to seventeeth year; studied medicine three years; never the slightest asthenopia. L.E., V. = $\frac{20}{20}$, all convex glasses blur.

Four grain solution atropia four times in eighteen hours. V. $\frac{20}{20}$ with + $\frac{1}{30}$, = $\frac{20}{20}$.

X.—D. M., æt. 26. Has been a student since five years of age. V. $\frac{20}{20}$; atropia in left eye for two days. V. = $\frac{20}{40}$ with + $\frac{1}{42}$ = $\frac{20}{20}$; third day $\frac{20}{30}$; with + $\frac{1}{30}$, $\frac{20}{20}$.

XI.—W. A. W., æt. 24. Always a student. V. = $\frac{20}{20}$; + $\frac{1}{80}$ blurs. Atropia in left eye for two days. V. = $\frac{20}{100}$; with + $\frac{1}{24}$, $\frac{20}{20}$ —(it is a dark day), and the observer's vision is also $\frac{20}{20}$.—

XII.—G. M. B., æt. 23. Went to school from five to eleven; then salesman and bookkeeper; has studied medicine two and a-half years. R.E., V = $\frac{20}{25}$; L.E., $\frac{20}{20}$; all convex glasses blur; atropia one day. R.E. V. = $\frac{20}{50}$; with + $\frac{1}{30}$, = $\frac{20}{20}$.

XIII.—B. W., æt, 27. Went to school four months a year until fifteen; then to an academy two years for eight months in a year; taught school four years; has been studying medicine for three years. V. = $\frac{20}{20}$ + $\frac{1}{12}$, blurs; atropia one day in each eye. V. = $\frac{20}{60}$, each eye, with + $\frac{1}{24}$, $\frac{20}{20}$.

XIV.—Dr. S., æt. 22. V. = $\frac{20}{20}$, each eye. Was for two weeks under the influence of a four-grain solution of sulphate of atropia; suffered from great dryness of the throat; atropia did not alter vision, and + $\frac{1}{80}$ blurred before and after its use.

To these fourteen cases I will add one of a lady thirty-two years of age, who consulted me on account of neuralgic pains about the eyeball and orbits. There was also asthenopia, and she complained of muscæ. She had been unable to do fine work for the past two months. The patient is a nervous lady who is somewhat overworked with household cares. The refraction seemed to be H. by the ophthalmoscope. V. = $\frac{20}{20}$, and all glasses blur; after atropia was used six days the vision was $\frac{20}{20}$, and the patient still rejected all glasses.

Summary.—Total number, 14.

Emmetropia, 3.

Hypermetropia, 11.

Degrees of H.:

$\frac{1}{36} \frown \frac{1c}{60}$ a 90°.

$\frac{1}{42}$ L.E., $\frac{1}{30}$.

$\frac{1c}{60}$, 90°.

R. E., $\frac{1}{30}$, and L.E., $\frac{1}{24}$.

$\frac{1}{36}$.

$\frac{1}{36}$.

$\frac{1}{30}$.

$\frac{1}{30}$.

$\frac{1}{24}$.

$\frac{1}{30}$.

$\frac{1}{24}$.

OBSERVATIONS UPON THE EFFECTS OF TOBACCO.*

By EDWARD T. ELY, M.D.

I HAVE recently had the opportunity of examining over one hundred workers in tobacco, chiefly cigar-makers. My primary object was to examine their eyes, but I proposed also to make a thorough examination of their general bodily condition. This general examination I was obliged to omit, because the investigations were made in the factories, and the men were unwilling to be taken from their work merely in the interests of science. It was by no means easy to obtain from them even the brief interviews with which I was favored.

The effects of tobacco upon the health must always have a practical interest for us, considering the millions who consume it and work in it, and the numerous questions regarding it which patients put to their medical advisers. I therefore venture to communicate to you the results embodied in this paper, feeling that, even if they be of no value for the branch of medicine for which they were collected, they may at least be a contribution to the hygiene of occupation.

Cigar-making is a sedentary and, upon all our theories, a very unwholesome pursuit. Those who have never been much in cigar-factories have no conception of their atmosphere. Ordinarily, a large number of workers are congregated in a single room, with all the windows closed, unless the weather is very mild. The impurity of the air, simply

---

* Read before the New York Clinical Society.

from overcrowding, is very oppressive to one unaccustomed
to it. There are added to this the emanations from large
quantities of tobacco and a never-ending cloud of tobacco-
smoke. It is not necessary to mention in detail the various
processes of *stripping*, *sweating*, *drying*, etc., by which the
tobacco is prepared for use, and by which it comes in contact with the operatives. Cigar-makers of the male sex
nearly all smoke. Of some it may be said that they smoke
almost incessantly. Many of the moderate smokers suck
the burnt stump of a cigar a large part of the time while
at their work. They are constantly handling tobacco, and
their fingers are more or less coated with it. Often there is
a thick crust of it upon their lips, from their habit of biting
off parts of the leaf instead of cutting them. Payment
by the piece is to some an incentive to lengthen their
hours of labor, so that they work at night as well as during
the day. Those whom I saw, being mostly Spaniards, seemed to be temperate, as a rule, drinking little besides red
wine. None of them chew tobacco. The only chewers in
the list are Americans.

The method which I followed was to test the vision of
each person, then to examine the eyes externally and with
the ophthalmoscope, then to ask a few general questions
about the health, habits, etc. The results, as I recorded
them in one hundred and two cases, are given in the following table. Some of the persons examined have been omitted from the list, because they were very young, and had
been only about a year in the business. All those excluded had perfect vision.

OBSERVATIONS UPON THE EFFECTS OF TOBACCO.

No.	Sex.	Age.	Nativity.	How many years work'd in Tobacco	Smoke. Chew. Drink.	Vision.	By Ophthalmoscope, etc.	General Remarks.
1	Male.	52	English.	40	S., 40 years, 10-15 cigars daily.	$\frac{20}{20}$	Normal.	"Doesn't hurt me a bit. Have seen thousands of cigar-makers; never knew one to give up work from failing sight." Looks well.
2	Male.	49	Spanish.	35	S., 35 years, to excess.	$\frac{20}{20}$ with $+\frac{1}{18}$ / 1 Jaeger w $+\frac{1}{8}$	Normal. H.	Feels perfectly well. Is fleshy. Some catarrh of pharynx.
3	Male.	49	English.	35	S., 30 years, average 3 daily.	$\frac{20}{30}$	Optic discs look atrophic.	"My sight has failed in past 6 or 7 years. My eyes water a great deal in the street." Is fleshy. Feels well.
4	Male.	45	Spanish.	29	S., 37 years, cigars and "a great many cigarettes."	1 Jaeger with difficulty.	Discs look pale; slight conjunctival catarrh.	"Tobacco doesn't hurt me." Looks and feels well.
5	Male.	43	German.	29	S., 18 years, 1-3 daily.	$\frac{20}{25}$	Patches of pigment at outer side of left disc. Disks pale at outer sides.	"Doesn't hurt me." Looks well.
6	Male.	40	......	26	S., 20 years, 1-2 daily now; more formerly.	R $\frac{20}{40}$ L $\frac{20}{20}$	Nerves pale. Has had divergent squint of right eye since 5 years old.	"It doesn't hurt me. My sight in the right eye is as good as it ever was."
7	Male.	38	Spanish.	25	S., 25 years. "I smoke all day."	$\frac{20}{20}$	Normal.	"I feel as well as I ever did."
8	Male.	38	Negro born in Cuba.	25	S., 10 years, excessively.	$\frac{20}{20}$	Discs pale. H.	"Doesn't hurt me. I feel better than when I began to work."
9	Male.	37	Spanish.	25	S., 25 years, 2-3 daily average; some cigarettes.	$\frac{20}{20}$ w $-\frac{1}{16}$ c 20	Normal.	Feels well except some dyspepsia. Has heard of "consumption" among cigar-makers, but never of failure of vision.
10	Male.	39	American.	25	S., 28 years, excessively; 6-8 daily now.	$\frac{20}{20}$	Normal.	Thinks tobacco doesn't harm him. Looks well.

OBSERVATIONS UPON THE EFFECTS OF TOBACCO.

No.	Sex.	Age.	Nativity.	How many years work'd in Tobacco.	Smoke. Chew. Drink.	Vision.	By Ophthalmoscope, etc.	General remarks.
11	Male.	37	Spanish.	24	S., 24 years, very much," "Not	$\frac{20}{50}$	Nothing special.	Dyspepsia 14 years. Pains in occiput. Nervous. Under care of a physician now. "I don't see well in the latter part of the day. I drink everything,—brandy, whiskey, everything." Breath alcoholic.
12	Male.	35	German.	22	S., 22 years, 2-3 daily.	$\frac{20}{20}$	Normal.	"It doesn't hurt me."
13	Male.	35	German.	22	S., 17 years. "I smoke all the time when not working; 4-5, a day during work."	$\frac{20}{20}$	Normal.	Nausea sometimes from smoking.
14	Male.	36	Spanish.	21	S., 20 years, 4-5 daily: often more.	$\frac{20}{20}$	Normal.	"Dyspepsia if I smoke too much." Well otherwise. "Never heard cigarmakers complain of failure of vision. Men work to age of fifty-five or sixty, as a rule."
15	Male.	39	Spanish.	20	S., 20 years.	$\frac{20}{20}$	Nothing special.	"It doesn't hurt me any."
16	Male.	53	Chinese.	20	S., 30 years, 2-3 daily.	$\frac{20}{20}$	Healthy. Very small pupils.	Is an opium smoker.
17	Male.	37	English.	20	S., 20 years, 5-6 daily.	$\frac{20}{20}$	Normal.	"Pain over heart sometimes." Otherwise well.
18	Male.	36	German.	20	S., over 20 years, 2-3 daily average.	$\frac{20}{20}$	Normal.	"Doesn't hurt me."
19	Male.	38	German.	20	S., 22 years, 3-4 daily.	$\frac{20}{20}$	Discs look pale.	"Doesn't hurt me."
20	Male.	29	German.	18	S., 6 years.	$\frac{20}{20}$	Normal.	"It gives me chronic catarrh."
21	Male.	30	German.	18	S., 18 years, 2-3 daily.	$\frac{20}{20}$	Normal.	Well.
22	Male.	28	Spanish.	18	Never has smoked.	$\frac{20}{20}$	Normal.	Well.
23	Male.	36	Spanish.	18	S., 25 years, 12 daily average now.	$\frac{20}{20}$	H.	"I am perfectly well."

						Vision		
24	Male.	34	Spanish.	17	S., 22 y'rs. "A great many cigarettes."	$\frac{20}{20}$	Normal.	. . . .
25	Female.	26	Bohemian.	17	. . . .	$\frac{20}{20}$	Normal.	. . . .
26	Male.	28	American.	17	S., 15 years, average 4 daily. "Sometimes 10."	$\frac{20}{20}$	Normal.	Feels well.
27	Male.	33	Spanish.	16	S., 16 years.	$\frac{20}{20}$	Normal.	"It doesn't hurt me."
28	Male.	32	Spanish.	16	S., 19 years, 12 cigars daily. "About 50 cigarettes daily."	$\frac{20}{20}$	Normal.	"It doesn't hurt me." Feels well.
29	Male.	37	Spanish.	16	S., 25 years. "I smoke all the time." Thinks he has smoked 30 cigars a day in Cuba.	$\frac{25}{20}$	Normal.	Feels perfectly well. Is fleshy. Never heard of tobacco injuring sight of workmen. Thinks it injures the lungs.
30	Male.	26	English.	16	S., 16 years, 2–3 daily.	$\frac{20}{20}$	Discs look red.	"Makes me nervous and spoils my appetite."
31	Male.	29	American.	16	S., 13 years, 2–3 daily.	$\frac{20}{20}$	Normal.	"Never heard of injury of sight from tobacco."
32	Male.	28	Spanish.	15	S., 13 years, 4–5 daily.	$\frac{20}{20}$	Normal.	"Tobacco doesn't hurt me." Looks well.
33	Male.	30	Spanish.	15	S., 13 years, 3–4 daily.	$\frac{40}{20} - w - \frac{1}{15} \text{ c } 180°$	Mixed astigmatism.	Has dyspepsia. Looks pale. Thinks his vision has failed since he was a boy.
34	Male.	25	Spanish.	15	S., 16 years, "I smoke all day." Apparently smokes only 3 or 4.	$\frac{20}{20}$	Normal.	"It doesn't hurt me."
35	Male.	23	Spanish.	15	S., 10 years, 5–6 daily.	$\frac{20}{20}$	H.	"Doesn't hurt me." Looks well.
36	Male.	29	German.	15	S., 15 years, 2–3 daily.	$\frac{20}{20} - w - \frac{1}{48} \text{ c } 180°$	Hyperopic astigmatism.	"Doesn't hurt me. My sight is as good as ever."
37	Male.	38	Bohemian.	15	Never has smoked.	$R \frac{20}{20} w + \frac{1}{36}$ $L \frac{20}{20} - w + \frac{1}{10}$	High degree of H. Strong convergent squint of right eye in trying to see letters at 20 feet.	. . . .
38	Male.	32	American.	15	S., 10 years, 3–4 daily 6 years; 10 daily past 4 years; "Some days I smoke 25."	$\frac{20}{20}$	Not examined.	Has seen thousands of cigar-makers. Never heard them complain of failing vision. "I consider it a healthy occupation."
39	Male.	27	Spanish.	14	S., 17 years, "I smoke all the time—20 cigars a day."	$\frac{40}{20}$	Normal.	"Doesn't hurt me." Has gained flesh. Never heard workmen complain of eyesight.

OBSERVATIONS UPON THE EFFECTS OF TOBACCO.

No.	Sex.	Age.	Nativity.	How many years work'd in Tobacco.	Smoke. Chew. Drink.	Vision.	By Ophthalmoscope, etc.	General remarks.
40	Male.	29	Spanish.	14	S., 10 years, 3–4 daily.	$\frac{20}{20}$	H.	Feels well. Looks well.
41	Male.	27	Spanish.	14	S., 12 years, 3–4 daily; sometimes more.	$\frac{20}{20}$	H. Jr.	"I feel as well as ever." Bad teeth. Has asthenopia.
42	Male.	43	Chinese.	14	S., 30 years, 10 daily.	$\frac{20}{30}$	Normal. Small pupils.	"I smoke a quarter of a pound of opium a week. Tobacco doesn't hurt me." "It doesn't hurt me."
43	Male.	49	Spanish.	14	S., 25 years, 6 daily.	$\frac{20}{20}$	Normal.	
44	Male.	37	Swedish.	14	S., 19 years, 2–3 daily.	$\frac{20}{70} - w - 2\frac{1}{2}$	Myopia. Post. staphyloma.	"Always near-sighted. Tobacco don't hurt me."
45	Male.	43	Spanish.	13	S., 30 years, 2–13 daily.	$\frac{20}{20} - w - \frac{1}{36}$ c 180°	Normal.	Is thin. Has dyspepsia. Looks well. Pharyngitis.
46	Male.	30	Spanish.	12	S., 18 years, 12 daily.	$\frac{20}{20}$	Normal.	Chronic cough Bad teeth.
47	Male.	25	American.	12	S., 8 years, 3–4 daily.	$\frac{20}{20}$	Discs pale.	Feels well.
48	Male.	26	Chinese.	12	S., 10 years, 1–3 daily.	$\frac{20}{20}$	Normal.	Opium-smoker.
49	Male.	33	Spanish.	12	S., 12 years, 6 daily.	$\frac{20}{20}$	Normal.	"It doesn't hurt me."
50	Male.	26	Spanish.	12	S., 12 years, 3–4 daily.	$\frac{20}{20} w + \frac{1}{36}$ c	H. As.	"It doesn't hurt me."
51	Male.	34	Swedish.	12	S., 16 years, 2 daily.	R $\frac{20}{200}$ L $\frac{20}{50}$	Chronic conjunctivitis. Extensive corneal opacities.	"Tobacco doesn't hurt me."
52	Male.	26	Negro born in Cuba.	11	S., since a boy excessively.	$\frac{20}{20}$	Normal.	"It doesn't hurt me." Has a chronic cough.
53	Male.	48	German.	11	S., 32 years, pipes and cigars to excess for 22 years; 10 cigars a day past ten years.	$20 \, w - \frac{1}{12}$	Not examined.	Feels perfectly well.
54	Female.	30	Bohemian.	11	....	$\frac{20}{20}$	Outer part discs pale.	Is well.
55	Male.	27	French.	11	S., 12 years, 3–4 daily.	$\frac{20}{20}$	H. Large physiological excavation.	"It gives me dyspepsia."

56	Male.	34	American.	11	S., 16 years, 6 daily 10 years; 15 daily past 6 years, often more; chews.	$\frac{30}{20}$	Normal.	Feels perfectly well. Never been ill.
57	Male.	26	Spanish.	10	S., 10 years. "Moderately."	$\frac{20}{20}$ w $-\frac{1}{30}$ c 180°	Normal.	"Sometimes makes me nervous. I can't write a letter at night because my hands shake." Is well.
58	Male.	26	Spanish.	10	S., 8 years, a good many cigarettes, as well as cigars.	R $\frac{20}{40}$ / L $\frac{20}{20}$ w $+\frac{1}{60}$ c 90°	Iritis of right eye and spots on lens. Left eye normal.	Is well.
59	Female.	..	Bohemian.	10	....	$\frac{20}{30}$+ w $+\frac{1}{36}$ c 90°	Normal.	Is well.
60	Male.	26	Spanish.	9	S., 10 years. "I smoke all day."	$\frac{20}{20}$	Normal.	Feels well. Catarrh of pharynx.
61	Male.	29	Spanish.	9	S., 12 years.	$\frac{20}{20}$	Normal.	"Nervous sometimes."
62	Male.	25	Spanish.	9	S., 10 years, 2–3 daily.	$\frac{20}{20}$	H.	"It doesn't hurt me." Looks well.
63	Male.	23	Spanish.	9	S., 9 years, 3–4 daily.	$\frac{20}{20}$	Normal.	"Do not feel as well as when I began. Felt *stronger* then; now more *lively.*" Is thin.
64	Male.	26	Spanish.	9	S., 9 years.	$\frac{20}{20}$	Outer part of discs pale. H.	"Makes me cough."
65	Male.	31	Chinese.	9	S., 5 years. "Not much."	$\frac{20}{20}$		....
66	Male.	..	French.	9	S., 20 years, 3–4 daily.	$\frac{20}{20}$	Outer part of left disc pale. Normal.	"Doesn't hurt." Is very deaf. Severe catarrh.
67	Male.	23	German.	9	S., 7 years, 3–4 daily.	$\frac{20}{20}$	H.	"Doesn't hurt me."
68	Male.	30	Spanish.	9	S., 17 years, 3–4 daily.	$\frac{20}{20}$	H.	"Doesn't hurt me."
69	Male.	25	German.	9	S., 15 years, 4–5 daily.	$\frac{20}{20}$ w $-\frac{1}{4}$ H c 180°	H. As.	"Doesn't hurt me. Sight is as good as ever."
70	Male.	23	Spanish.	.8	S., 8 years. "A good deal."	$\frac{20}{20}$	Normal.	"I don't feel as well as when I began."
71	Male.	23	Spanish.	8	S., 6 years, 3–4 daily.	$\frac{20}{20}$	H. b. Nothing else.	"My sight has always been poor."
72	Male.	47	Spanish.	8	S., 30 years, 4–6 daily.	$\frac{20}{20}$	Normal.	"Doesn't hurt me.'
73	Male.	31	German.	8	S., 20 years, 2–4 daily.	$\frac{20}{20}$	H.	"Doesn't hurt me."
74	Female.	25	Bohemian.	8	....	$\frac{20}{20}$	Normal.	Is well.

OBSERVATIONS UPON THE EFFECTS OF TOBACCO.

No.	Sex.	Age.	Nativity.	How many years work'd in Tobacco.	Smoke. Chew. Drink.	Vision.	By Ophthalmoscope. etc.	General Remarks.
75	Male.	25	American.	8	S., 16 years. "A great deal." 7–8 daily now.	$\frac{20}{20}$	Normal.	Has dyspepsia.
76	Male.	23	Spanish.	7	S. 7 years, to excess.	$\frac{20}{20}$	Normal.	Feels well. Looks well. Large tonsils. Pharyngitis. Bad teeth. "Doesn't hurt me."
77	Male.	30	Spanish.	7	S., 12 years, to excess.	$\frac{20}{30}-$	Discs pale. H.	"Doesn't hurt me."
78	Female.	19	Bohemian.	7	....	$\frac{20}{20}-$	Normal.	Is well
79	Male.	23	Spanish.	7	S., 7 years, 5–6 daily.	R $\frac{20}{20}- w+1\frac{1}{2}$ L $\frac{20}{100} w+1\frac{1}{4}$	H. ½ Nothing else.	"Doesn't hurt me."
80	Male.	20	Spanish.	7	S., 5 years, 3–4 daily.	$\frac{20}{20}$	H.	"Doesn't hurt me."
81	Male.	21	Spanish.	6	S., 7 years.	$\frac{20}{20}$	Normal.	Feels well as ever.
82	Male.	21	Spanish.	6	S., 7 years. "A good deal."	$\frac{20}{30}+$	Normal.	"Doesn't hurt me."
83	Male.	22	Spanish.	6	S., 6 years, 4–5 daily.	$\frac{20}{70}+ w+1\frac{1}{2}$	....	No evidence that sight has failed in past 6 years.
84	Male.	32	Chinese.	6	S., 9 years, 5–6 daily.	$\frac{20}{20}$	Normal.	....
85	Male.	29	German.	6	S., 15 years, 2–5 daily.	$\frac{20}{20}$	Normal.	"Doesn't hurt me."
86	Female.	16	Bohemian.	6	....	$\frac{20}{20}-$	H. Large retinal veins.	Is well.
87	Female.	27	Bohemian.	6	....	$\frac{20}{20}$	H.	....
88	Male.	20	Spanish.	5	S., 5 years.	R $\frac{20}{20} w-\frac{1}{24} c\,130°$ L $\frac{20}{40}$	Right eye H.; left eye M. As. Nothing else.	Dyspepsia and pains in back. Otherwise well. Looks well. Bad teeth. "My left eye was sore when young."
89	Male.	17	American.	5	S., 5 years.	$\frac{20}{20}$	Normal.	Is well. Some dyspepsia.

No.	Sex	Age	Nationality		Habit	Vision	Normal. Slight Blepharitis.	Remarks
90	Male.	23	American.	5	S., 8 years, 10 daily.		Normal. Slight Blepharitis.	Feels well. Looks well.
91	Male.	26	Swiss.	5	S., 8 years.	$\frac{20}{20}$ -w -$\frac{1}{48}$ c	H. As.	"Doesn't hurt me."
92	Male.	19	American.	5	S., 9 years, to excess; 5 daily now; chews a good deal.	$\frac{20}{20}$	H.	Perfectly well.
93	Male.	19	Bohemian.	5	S., 7 years. "Not much."	$\frac{20}{20}$	H.	. . . .
94	Male.	35	Spanish.	5	S., 23 years, 3-4 daily.	$\frac{20}{20}$ -w +$\frac{1}{60}$ c 90° [-$\frac{1}{48}$ c	H. As. Large retinal veins.	Wears glasses for reading. Feels well. Some dyspepsia.
95	Male.	22	Spanish.	5	S., 10 years, 5-6 daily.	$\frac{20}{20}$	. . . .	"Doesn't hurt me."
96	Male.	22	American.	4	S., 8 years, 1-5 daily.	$\frac{20}{20}$	H.	Is well.
97	Male.	30	Spanish.	3	S., 17 years, 12 daily, and more for past 3 years.	$\frac{20}{20}$	Normal.	Feels well.
98	Female.	18	Bohemian.	3	. . . .	$\frac{30}{20}$	H.	Is well.
99	Male.	23	Polish.	3	S., 3 years.	$\frac{20}{20}$	Normal.	. . . .
100	Female.	17	Bohemian.	2	. . . .	$\frac{20}{20}$	Normal.	Is well.
101	Male.	22	Spanish.	1½	S., 4 years, 4-5 daily.	$\frac{20}{20}$	H.	Is well.
102	Male.	80	American.	. . .	. . . .		. . . .	Asthenopia in evening work. "I feel weaker in every way than before going into the business." No dyspepsia. Has lost flesh. Looks well. Been a manufacturer 30 years. "Have seen thousands of workmen, and never heard of their sight being impaired by tobacco."

It seems to me that, among persons so exposed as these are to the influence of tobacco, by handling, smoking, and inhaling it for many hours each day, even one hundred individuals ought to show considerable of its bad effect, if this is very great. The general impression which I received from looking at the workmen as a whole was that of their being in average health. I got the same impression from visiting a large number of factories for cigars and cigarettes in Havana a few years ago.

This corresponds with observations reported by others, so far as male adults are concerned. Dr. Roger S. Tracy, who, with Dr. N. B. Emerson, has investigated the condition of cigar-makers in the factories and tenement-houses of this city, has expressed the opinion that the business is specially injurious only to persons under the age of puberty and to females. Dr. Tracy thinks sexual development is hindered in young girls by tobacco, and he and Dr. Emerson were greatly struck by the paucity of children in the families of cigar-makers. Thus, in one hundred and twenty-four families only an average of 1.09 children to each married couple was found; and in two hundred and one families only an average of 1·63 children.* This is surely a low ratio for a tenement-house population.

I tried to collect as much *oral* evidence as possible by questioning those workmen who seemed specially intelligent. The prevailing belief among them seemed to be that the injurious effects of tobacco consisted in " cough," " dyspepsia," and " nervousness," manifested in a limited number of persons. I examined the throat and teeth as often as possible. Pharyngeal catarrh was found very commonly, but perhaps no oftener than in other classes. The proportion of bad teeth did not seem large to me.

With regard to the effects of tobacco upon the vision, different authorities hold different opinions. From some articles upon the subject—such, for instance, as the valuable papers of Mr. Jonathan Hutchinson, of London—the reader might infer that impaired vision from tobacco was frequent (in England, at least), and that it was differentiated from

---

* Board of Health Reports, 1874-5.

other forms of amblyopia without much difficulty. From
the writings of some others the inference might be drawn
that it was an open question whether tobacco had any such
bad influence whatever.

Soelberg Wells, in his *Treatise on the Eye*,* says: "In
by far the greater number of cases of amaurosis which I
have met with in heavy smokers, the patients readily admitted their free indulgence in other excesses. I fully admit
the fact that the excessive use of tobacco (but most frequently together with other causes) may produce considerable impairment of vision, and finally, if the habits of the
patient be not entirely changed, and the use of tobacco,
stimulants, etc., given up, even atrophy of the optic nerves.
But I can not, from my own experience, accede to the doc-
trine that there is anything peculiar in the form of atrophy
of the optic nerve which would at once enable one to diag-
nose the nature of the disease as depending upon excessive
smoking."

Mr. Carter, in his text-book on the eye, in discussing the
causes of optic-nerve atrophy, writes as follows: "Among
those most commonly assigned, tobacco and alcohol held
prominent places, but held them, I venture to think, upon
very feeble and insufficient evidence," He then refers to a
statement by Dr. Dickson (physician to the British Embassy
at Constantinople), that the "consumption of tobacco in
that city averaged about three pounds weight per head per
month for the whole population, but that 'amaurosis' was
a 'rare affection' there." He also quotes a letter from Dr.
Hubsch, an oculist of Constantinople, as follows: "With
regard to the action of tobacco upon the eyes, it is very
problematical; here everybody smokes from eve to morn,
and from morn to eve; the men smoke much, the women a
little less, and children smoke from the age of seven or
eight. I have never been able to attribute amaurosis to the
abuse of tobacco; the number of smokers is immense, and
the number of amaurotic persons limited." † Mr. Carter
goes on to say: "I have obtained the same kind of nega-

---

* Page 450. American edition.
† Translated from the French. Carter Am. ed., p. 477.

tive evidence from Egypt and India : and, in the face of it, taking into account the difficulty of distinguishing between causation and coincidence, I do not attach much importance to the fact that several patients who have suffered from nerve atrophy have been 'great smokers' * * * and it is to my mind conclusive," Mr. Carter says, " that, although the consumption of tobacco has greatly increased of late years, I have no experience of any parallel increase of nerve atrophy." To this passage Dr. John Green, of St. Louis (who edits the American edition of Mr. Carter's book), appends a foot-note, in which he says: "A very large proportion of the cases of optic-nerve atrophy which have fallen under our own observation have been in cigar-makers and workmen in tobacco factories. Such persons ordinarily use tobacco very freely, both by smoking and chewing, but are not especially addicted to drinking."

Of the cases from the books of the Manhattan Eye and Ear Hospital, given below, not a single one was a worker in tobacco, and there was only one among the cases reported by Mr. Hutchinson. In the work of Stellwag greater stress is laid upon tobacco and alcohol combined as a cause of amblyopia than on tobacco alone. In a recent article * upon "Tobacco and Alcohol Amblyopia," by Dr. T. Hirschberg, of Berlin, the author expresses his decided belief in a characteristic form of amblyopia due to the abuse of tobacco alone.

De Wecker, of Paris, whose text-book is one of the latest, says: "The most commonly met with form of toxic amblyopia is that produced by the conjoint abuse of alcohol and tobacco. Some authors are still doubtful whether tobacco alone ever produces it. Before going further, I must insist on the importance of your satisfying yourselves by ophthalmoscopic examination, and in other ways, that no errors of refraction exist uncorrected. Such might readily produce amblyopia; indeed, I have been before now surprised to hear the opinions of men often cited in these cases who absolutely neglect this important preliminary examina-

---

* *British Med. Journal*, 1879, ii, p. 810.

tion." * In recording his own experience, M. De Wecker speaks of "*tobacco and alcohol combined*" as causing the amblyopia.

These brief quotations will be enough to represent to you the prevailing opinions upon this subject. Those who believe most firmly in tobacco amblyopia consider that the prognosis is good if the bad habit be abandoned, and that, under these circumstances, atrophy of the optic nerve is a rare sequel.

A tobacco amblyopia is mentioned in all modern textbooks on the eye, and most ophthalmologists believe in it, and on good and sufficient evidence, so far as we can judge. It certainly seems to me to have an influence in causing impairment of the sight, but to a less extent than some writers would have us believe. In quite a number of cases of so-called *amblyopia from abuse*, seen by me in the past few years, there have been but few in which the abuse of tobacco was not combined with other abuses, especially with that of alcohol.

In the records of Dr. Roosa's practice and my own for the past few years I find twenty-one cases which were regarded as amblyopia from abuse. Of these, thirteen persons admitted excessive indulgence in both alcohol and tobacco; two used tobacco to excess, and said they used liquor only "moderately;" five admitted excess only in tobacco; and one abused alcohol alone. Of those who used tobacco excessively, it is recorded that one had syphilis, and that one had been "straining his eyes in doing some very fine inlaid wood-work" when the amblyopia came on. A man is now under observation who has been a great drinker and smoker for ten years, and who has noticed failing vision for the past six months. His vision is $R.\ E.$, $\frac{3}{200}$; $L.\ E.$, $\frac{10}{200}$; and he has neuro-retinitis in both eyes. There is no proof that the inflammation is dependent upon his excesses, although no other cause is apparent; if, however, he had presented himself afterward with atrophied nerves, his condition would undoubtedly have been ascribed to his bad habits.

---

* *Ocular Therapeutics.* Translated by Litton Forbes. London, 1879, p. 446.

Dr. Stowell has kindly looked over the records of the Manhattan Eye and Ear Hospital for the past seven years, with the following result: forty cases of amblyopia, with or without atrophy of the optic nerves, are recorded as associated with the abuse of spirits and tobacco. Of these, eight persons are recorded as abusing tobacco alone; twenty-two, liquor alone; ten, both combined. As already mentioned, there is no worker in tobacco in the list, but there are three liquor-dealers. The occupations were given as follows: laborers, ten; boatmen, two; carpenters, four; foreman, one; machinists, two; liquor-dealers, three; drivers, three; clerks, three; soldier, one; printers, two; tailor, one; physician, one; stableman, one; farmer, one; artist, one; marble-cutter, one; waiter, one; longshoreman, one; housekeeper, one. One of the patients was a female (a housekeeper), of whom it is recorded that she had "long been in the habit of drinking and smoking." She had atrophy of the optic nerves with vision equal to counting fingers at four feet, each eye.

Mr. Hutchinson says,[*] "Total abstainers from stimulants are more liable to suffer than others, and, although we sometimes meet with the disease in the intemperate, I have a strong impression that, on the whole, alcohol counteracts tobacco." I do not understand this to be the general opinion among ophthalmologists, but rather the reverse. In at least half of the cases reported by Mr. Hutchinson the patients were drinkers. In commenting on a series reported in 1871, he says, "In all the worst cases the patients had used alcoholic drinks, and two of them had been great drinkers." [†]

What De Wecker says about examining the refraction, etc., is very important. Doubtless much error has arisen here, as elsewhere, through careless observation. Great care is needed in taking the history of such cases as are here referred to; for everybody of experience knows what a rigid cross-examination is required to get from a patient any true account of a failure of one of the special senses. In the

---

[*] " Royal London Oph. Hosp. Reports," 1876, p. 458.
[†] " Royal London Oph. Hosp. Reports," 1871. p. 185.

hurried examinations of dispensary practice (and sometimes of office practice) a person will often give an apparently clear history of failing sight from the abuse of tobacco, when more rigid inquiries will fail to show that the vision is any worse than it was before the bad habit was contracted. It will be found, for example, that the patient reads the same print that he always has read, recognizes distant objects as well as ever; that, in short, there is no evidence of any failure of vision, but only of more attention than usual being paid to a defect which has always existed.

The making of cigars requires good eyesight, as part of the work is quite fine. Especially is this true of the small end of the cigar. The making of this point, and the closing of it accurately, so as to leave no seam or gap visible, is a matter of pride among good workmen. If impairment of sight were a frequent or constant effect of tobacco, many cigar-makers would be rendered incapable of doing first-class work. Is it not reasonable to suppose that some of them would seek advice from doctors, and would be told that their bad vision was caused by tobacco; that the matter would be talked over at the work-bench; and that in time a certain tradition would grow up about it among the operatives themselves?

In my inquiries among the most intelligent workmen, before referred to, I did not find one, even of those grown old in the business, who had ever heard of vision being impaired by tobacco. Of course, no great stress can be laid upon this kind of evidence, but still it has a certain value. I found several men with presbyopia and hypermetropia, who were embarrassed in doing the finer parts of their work, and who might have passed for cases of amblyopia. Proper glasses, however, restored their vision at once to the normal standard. One man, for instance, a Spaniard, forty-nine years old, who had worked at tobacco and smoked for thirty-five years, needed convex glasses of eighteen inches focus to give him normal distant vision, and glasses of eight inches focus for near objects. He was working with difficulty, and complaining of failing sight, without any idea of the real cause of his defect. I was told that old cigar-makers scarcely ever wore glasses at their work.

It will be seen by the table that 88 of the subjects had what may be called normal vision. Where the vision is marked $\frac{20}{20}$—, it means that the subject miscalled one or two of the test-letters, but no more than was justly attributable to ignorance, excitement, or a refractive defect. This is probably as good a showing as would be obtained from 100 persons taken at random anywhere. Of the 15 who had defective vision in one or both eyes, in 13, it seems to me, the amblyopia is explainable by the refractive condition, or by the history of the case. At any rate, there is no good reason for attributing it to tobacco. In case No. 11 the patient probably had syphilis. There remain only two cases, in my opinion, in which the amblyopia can be fairly attributed to tobacco, provided one chooses to do so. These are Nos. 3 and 77. Of course, even in those cases, there is no positive proof one way or the other.

No importance is attached to the variations in color of the optic discs, noted in many cases, for they were only such as are constantly seen in non-smokers, such as women and children, and in other eyes, which, for all practical purposes, we are obliged to consider healthy. The visual fields ought to have been tested in these examinations, but it was impossible to do so. The vision was tested under uniform illumination, as far as possible, and the tests were varied enough to prevent deception.

It will be seen by the table that 59 of the subjects (or about 60 per cent.) had been working in tobacco for upward of 10 years. Of these, one had worked 40, two 35, two 29, one 26, four 25, one 24, two 22, one 21, five 20, four 18, three 17, five 16, seven 15, six 14, and six 12 years.

Twelve were 40 years old or more, and forty-seven were 30 or more.

There were 48 Spaniards, 13 Germans, 12 Americans, 11 Bohemians, 5 Chinese, 4 English, 2 French, 2 negroes, 2 Swedes, 1 Swiss, and 1 Pole. The list includes nine females, all of whom were in good health and had normal vision.

Of the 93 males, 63 had been smokers for ten years or more. Some had been excessive smokers for half of a lifetime. Two had never smoked at all. Of course, it was not possible to obtain accurate information as to the amount of tobacco used by each smoker. Undoubtedly some exaggerated the amount, while others understated it.

My own impressions, gathered from these examinations, as well as from other experience, are, that tobacco has of itself only a comparatively slight influence in impairing the vision ; that working in tobacco is as healthful as most other sedentary occupations; that in certain persons peculiarly susceptible to it, or, when combined with other noxious influences, it may impair the vision or the general health, just as has been claimed for it ; and that constant contact with it, as with other poisons, may beget a tolerance of it sufficient to contradict all theory.

The strength of the weed and the manner of using it have an influence, and this may explain the freedom from tobacco-amblyopia which certain countries are said to enjoy.

The conclusions drawn here are not new, but it may not be unprofitable to present them afresh for your consideration.

My thanks are due to Mr. Howard Ives and to R. Monné & Brother, of this city, for the privilege of making these examinations.

NOTE.—In the table, the expression $\frac{20}{20}$ denotes perfect vision. When it is followed by the *minus* sign ($\frac{20}{20}$ —) it means that the vision was a trifle below the normal standard. The other fractions denote various degrees of impaired vision and the glasses with which they could be wholly or partially corrected. The letters H., M., As., signify the refractive defects, hypermetropia, myopia, and astigmatism.

ADDENDUM.

A CASE OF SUPPOSED AMBLYOPIA FROM QUININE POISONING REPORTED ON PAGE 46.

When the report of the case, on page 46, was already in press, we found that Dr. A. H. Voorhies, of Memphis, Tenn., and Dr. L. De Wecker, of Paris, had recently made valuable contributions to this subject.

The following extracts are from the report of a case published by Dr. Voorhies in the *Transactions of the American Medical Association* for 1879, p. 411 :

"February 16, 1878, I was asked to see Miss V. H., a young lady, aged 18, living on the Arkansas side of the Mississippi. * * * I found the patient in bed, with every appearance of being extremely ill. * * * One week before the date of my visit, under the apprehension that this lady was threatened with a congestive chill, a relative (not a physician) caused an ounce of quinine to be administered to her within the space of a few hours, and that a like quantity was given each day for the two following days. In other words, more than 1,300 grains were given by stomach and rectum within three days. On the morning of the second day it was discovered that she was perfectly blind. * * * Audition was but slightly impaired. * * * There was marked paleness of the face, and this was also noticed in the conjunctival lining of the lids. Pupils normal, responding promptly to light. T : 1 ; anæsthesia of cornea so as to suffer a probe moved over its surface without complaint. No perception of light. Ophthalmoscopic view very peculiar ; disc perfectly white ; not a trace of optic nerve vessels—neither veins nor arteries ; choroidal vessels empty, with pale yellowish tinge of retina."

Nitrite of amyl was used for four days without any apparent effect. Strychnine was then given hypodermically. * * * "But still no improvement of sight was discovered until the middle of the tenth week, when she was able to discern a trace of light on the use of the ophthalmoscope. The return of light was very gradual, until she was, and is now, enabled to read Jaeger No. 1. I had the opportunity of examining this case a few days ago, and find the disk perfectly white, with still no trace of central artery except a small twig, which is just perceptible as it struggles over the upper half of the disk of the left eye, to be lost on reaching the retina. The field of vision is greatly contracted, and * * * I find the greatest, which is the vertical diameter, to be less than four inches when taken at two feet. Her general health is good, and her mental activity is fully restored."

The following is extracted from Dr. Wecker's recent work *

" Intoxication by *quinine* is extremely rare. You have seen here a young patient, who, having contracted intermittent fever in the tropics, determined to cure himself. He filled a large glass for about an inch with quinine, swallowed it all, and went to bed. He awoke both deaf and blind. Hearing and vision eventually returned, but the latter imperfectly. For though central acuity was normal, the visual field of each eye showed a peculiar symmetrical lacuna. There were, in both visual fields, islands of blindness ; the larger of the two occupied a considerable portion of the internal half of the field, and extended somewhat beyond the point of fixation ; the smaller affected merely a small external portion. In these rare cases, so far as my experience goes, vision has returned incompletely, and I have never seen absolute blindness caused. * * * "

---

* Ocular Therapeutics. By L. De Wecker. Translated by Litton Forbes, M.D. London, 1879.

www.ingramcontent.com/pod-product-compliance
Lightning Source LLC
Chambersburg PA
CBHW021941190326
41519CB00009B/1098